Elastic and Elastoplastic Contact Analysis

Using Boundary Elements and Mathematical Programming

Topics in Engineering

This series provides a rapid and informal dissemination of significant new work in engineering. It is aimed at high level coverage across a broad field of engineering including mechanical, civil, hydraulic and structural, as well as other associated topics.

G. Schmid
Ruhr-Universität Bochum
Germany

M. Tanaka
Shinshu University
Japan

J.C.F. Telles
University of Rio de Janeiro
Brazil

H. Tottenham
Tottenham & Bennett, Consulting Engineers
UK

J.R. Whiteman
Brunel University
UK

Dedication

In memory of my dear father,

Hajj Amir Faraji (1939-2001)

Elastic and Elastoplastic Contact Analysis

Using Boundary Elements and Mathematical Programming

A. Faraji
University of Tehran, Iran

WIT*PRESS* Southampton, Boston

Elastic and Elastoplastic Contact Analysis

Using Boundary Elements and Mathematical Programming

Series: Topics in Engineering

A. Faraji

University of Tehran, Iran

Published by

WIT Press
Ashurst Lodge, Ashurst, Southampton, SO40 7AA, UK
Tel: 44 (0) 238 029 3223; Fax: 44 (0) 238 029 2853
E-Mail: witpress@witpress.com
http://www.witpress.com

For USA, Canada and Mexico

WIT Press
25 Bridge Street, Billerica, MA 01821, USA
Tel: 978 667 5841; Fax: 978 667 7582
E-Mail: infousa@witpress.com
http://www.witpress.com

British Library Cataloguing-in-Publication Data

A Catalogue record for this book is available
from the British Library

ISBN: 1-85312-733-7
ISSN: 0952-5300

Library of Congress Catalog Card Number: 2004113019

Printed in Great Britain by Cambridge Printing.

Contents

Preface

This book presents a general elastic and elastoplastic analysis method for the treatment of two- and three-dimensional contact problems between two deformable bodies undergoing small displacements with and without friction using the Boundary Element Method (BEM) and Mathematical Programming (MP). This is applied to the contacting bodies pressed together by a normal force which is sufficient to expand the size of the contact area, mainly through plastic deformation, acted on subsequently by a tangential force less than that necessary to cause overall sliding.

In order to analyze elastic frictional contact problems, following concepts borrowed from the theory of elastoplasticity, the contact separation-slip (yield) functions corresponding to the unilateral normal contact and to the sliding threshold contact laws are derived and using a penalty formulation, the contact problem is reformulated into a penalty complementarity form. For numerical discretization purpose, the boundary element formulation is performed and yields a discrete system of equations relating contact boundary tractions and displacements (contact variables) to the prescribed surface tractions and displacements. These relations are then coupled with the contact conditions through a penalty formulation. Finally, the coupled formulation is expressed as a mathematical programming problem, *i.e.* a linear complementarity problem (LCP) and solved by Lemke's algorithm. This method can avoid the tedious and costly classical iterative procedures and, if there is no unloading, only one incremental step is required to obtain a solution. This algorithm has been implemented in a computer code CONTACTA programmed in Fortran. Size of the contact patch, of the resulting zone of micro-slip in the frictional case, normal and tangential tractions distributions in the contact zone as well as stresses and displacements for each node of contacting bodies may be evaluated through this package. A validation of the numerical results is illustrated and compared with available published results.

In order to analyze elastoplastic frictional contact problems, assuming a linear elastic perfectly plastic material behavior and introducing the contact separation-slip (yield) and plastic yield functions, both the contact criterion and elastoplastic constitutive equations are expressed as a kind of linear complementarity problem (LCP). Based on these linear complementarity problems (LCPs), and on a boundary element formulation that utilizes some particular integrals (meshless method) to avoid volume integration, a general incremental boundary element-mathematical programming method is obtained in the form of a unified linear complementarity problem. Finally, the mathematical programming problem is solved by Lemke's algorithm and the analysis is successfully implemented through the computer package CONTACTA. Using this package, with the material is assumed to obey the von Mises or Tresca criteria with their associated flow rule, one can obtain the value of the load at the first yield for which the first node becomes plastic, as well as stress and displacement components for each node of the contacting bodies. When the load is increased beyond that level, the contact zone extension, distributions of normal and tangential contact tractions and resulting zone of micro-slip can be determined. Numerical results are given and compared with available results to demonstrate the efficiency and the accuracy of the method.

The author would like to express his sincere appreciation and deepest gratitude to Professor Alain Cardou and Professor Augustin Gakwaya (Department of Mechanical Engineering, Laval University, Québec, Canada) for their valuable advice, helpful comments, suggestions and discussions throughout the duration of this research. Special thanks are due to Professor Pierre Villon (Department of Mechanical Systems, Compiegne Technology University, Compiegne, France), Professor Roger Pierre (Department of Mathematics and Statistics, Laval University, Québec, Canada), Professor Sylvain Goudreau and Professor John Dickinson (Department of Mechanical Engineering, Laval University, Québec, Canada) who read the manuscript and helped to improve the text. The author wishes to thank his parents, his wife and children for their understanding, patience and encouragement. The financial support from the Ministry of Science, Research and Technology of I.R. Iran is greatly appreciated.

A. Faraji
2005

List of symbols

A_0	Characteristic length
b_{n}	Constant (can generally be set equal to unity)
$\beta_k = (k-1)\dfrac{2\pi}{m_{\mathrm{c}}}$	Direction of the normal vector to side k in the polygonal projection of the polyhedral approximation of the friction cone, where $k = 1, \ldots, m_{\mathrm{c}}$, $m_{\mathrm{c}} = 2$ in 2D and $m_{\mathrm{c}} = 8$ in 3D
$c^{\mathrm{i}}_{\ell k}$	Local geometric characteristics at the location i of the unit concentrated load
C	Number of nodes of the potential contact zone
$C(x, \xi_n)$	Global shape function
$[C_{\mathrm{c}}]$	Contact separation-slip (yield) matrix
$[C_\ell]$	Linearization matrix
$[C_{\mathrm{p}}]$	Plastic yield matrix
d	Number of dimensions of the model, i.e. $d = 2$ in 2D and $d = 3$ in 3D
$\mathrm{d}\varepsilon$	Volumetric incremental strain
$\mathrm{d}\varepsilon^{\mathrm{e}}_{ij}$	Recoverable incremental
$\mathrm{d}\varepsilon^{\mathrm{p}}_{ij}$	Plastic strain increment vector
$\mathrm{d}\varepsilon^{\mathrm{o}}_{ij}$	Permanent incremental inelastic (plastic) strain
$\mathrm{d}g_i$	Galerkin increment vector
$\mathrm{d}h_{ij}$	Tensor increment function
$\mathrm{d}\lambda_{\mathrm{p}}$	Plastic multiplier
$\{\mathrm{d}p_{\mathrm{c}}\}$	Incremental normal and tangential contact tractions
$\mathrm{d}\psi_{ij}(\xi_n)$	Fictitious increment tensor density
$\mathrm{d}\sigma_{ij}$	Incremental stress tensor
$\{\mathrm{d}\sigma\}$	Incremental stress
$\{\mathrm{d}\sigma^{\mathrm{c}}\}$	Complementary incremental stress vector
$\{\mathrm{d}\sigma^{\mathrm{p}}\}$	Particular incremental stress vector
$\mathrm{d}\sigma^{\mathrm{o}}_{ij}$	Initial incremental stress vector
$\{\mathrm{d}t\}$	Total incremental traction vector
$\{\mathrm{d}t^{\mathrm{c}}\}$	Complementary incremental traction vector
$\{\mathrm{d}t^{\mathrm{p}}\}$	Particular incremental traction vector
$\{\mathrm{d}u\}$	Total incremental displacement vector
$\{\mathrm{d}u^{\mathrm{c}}\}$	Complementary incremental displacement vector
$\{\mathrm{d}u^{\mathrm{p}}\}$	Particular incremental displacement vector

$\{du_c\}$	Incremental normal and tangential contact relative displacements
D^e_{ijkl}	Elastic constitutive tensor
$[D_c]$	Elastic (stiffness) matrix
$[D^s]$	Elastic constants matrix
$\Delta^i e_\ell$	Dirac delta function representing a unit load at the point i in the ℓ direction
δ_{kj}	Kronecker delta symbol ($\equiv 1$ for $k = j$ and $\equiv 0$ for $k \neq j$)
e_j	Eigenvectors of the symmetric stress tensor $[\sigma]$
ε_{kj}	Strain tensor
$\varepsilon_{\ell\ell}$	Volumetric strain
E	Young's modulus
E_n ($E_n \to \infty$)	Normal penalty parameter
E_t ($E_t \to \infty$)	Tangential penalty parameter
f_0	Yield limit in uniaxial loading ($f_0 = -Y$)
f_{c1}	Normal contact separation (yield) function
$f_{c(k+1)}$	Tangential contact slip (yield) functions, where $k = 1, \ldots, m_c$, $m_c = 2$ in 2D and $m_c = 8$ in 3D
$\{f_c\}$	Contact separation-slip (yield) functions
f_p	Plastic yield function
f_{pr}	Piecewise linear yield functions
$\{f\}$	Contact and plastic yield functions
$\begin{bmatrix} f_{oo} & f_{oc} \\ f_{co} & f_{cc} \end{bmatrix}^\alpha$	Flexibility matrix of two contacting bodies ($\alpha = 1, 2$)
$[\phi]$	Interpolation function matrix
g_{c1}	Flow potential function for normal contact problems
$g_{c(k+1)}$	Flow potential function for tangential contact problems, where $k = 1, \ldots, m_c$, $m_c = 2$ in 2D and $m_c = 8$ in 3D
$\{g_c\}$	Flow potential function for normal and tangential contact problems
g_p	Plastic flow potential function
$\dfrac{\partial g_p}{\partial \sigma_{ij}}$	Gradient vector
$G = \dfrac{E}{2(1+\nu)}$	Shear modulus
$[G]$	Influence matrix corresponding to the boundary displacements
$[G^\sigma]$	Influence matrix corresponding to the interior stress field
$[G^u]$	Influence matrix corresponding to the interior displacement field
$\gamma_r = (r-1)\dfrac{2\pi}{m_p}$	Direction of the normal vector to side r in the polygonal projection of the polyhedral approximation of the plastic yield function, where $m_p = 6$ for the Tresca and $m_p = 8$ for the von Mises yield condition
h	Hardening parameter
$[H]$	Influence matrix corresponding to the boundary displacements
$[H^\sigma]$	Influence matrix corresponding to the interior stress field
$[H^u]$	Influence matrix corresponding to the interior displacement field
$[I]$	Identity matrix
I_1, I_2 and I_3	Invariants associated with the stress tensor $[\sigma]$

J_1, J_2 and J_3	Invariants associated with the stress tensor $[\sigma']$
ξ_n	Boundary nodes along with a group of auxiliary of fictitious nodes in the domain
(ξ, η, ζ)	Local coordinate system on the potential contact surface $S_c (S_c = S_c^1 = S_c^2)$
L( )	Linear differential operator
λ_{c1}	Normal contact multiplier
$\lambda_{c(k+1)}$	Tangential contact multipliers, where $k = 1, \ldots, m_c$, $m_c = 2$ in 2D and $m_c = 8$ in 3D
$\{\lambda_c\}$	Normal and tangential contact multipliers
$\{\lambda\}$	Normal and tangential contact and plastic multipliers
Γ	Load factor
m_c	Number of the orientation of each plane in the piecewise linear approximation of the friction cone, in which $m_c = 2$ in 2D and $m_c = 8$ in 3D
m_p	Number of approximating yield functions (planes) or yield modes in the principal stress state, with $m_p = 6$ for the Tresca and $m_p = 8$ for the von Mises yield condition
ME	Number of nodes on an element
M	Number of surface (boundary) nodes
$[M]$	Known matrix
μ	Coefficient of friction
$(\mathbf{t}_1, \mathbf{t}_2, \mathbf{n})$	Unit basis vectors of local coordinate system (ξ, η, ζ)
$(\mathbf{t}_1^\alpha, \mathbf{t}_2^\alpha, \mathbf{n}^\alpha)$	Unit basis vectors of local coordinate system defined on the contact surfaces S_c^α of two contacting bodies $(\alpha = 1, 2)$
N	Sum of the number of surface (boundary) nodes and interior sample points
N_1, N_2	Sum of the number of surface (boundary) nodes and interior sample points of the two contacting bodies
NE	Number of elements
ν	Poisson's ratio
p	Number of independent stress components, i.e. $p = 3$ in 2D and $p = 6$ in 3D
$\{p_c\}$	Normal and tangential contact tractions
$(p_{cn}^\alpha, p_{ct_1}^\alpha, p_{ct_2}^\alpha)$	Normal and tangential contact traction components of two contacting bodies $(\alpha = 1, 2)$
$\{q\}$	Known vector
R^d $(d = 2, 3)$	Euclidean space
ρ	Euclidean distance between the field (sample) point and the source (load) point
$S_u(S_u^1, S_u^2)$	Surface where prescribed displacements are applied
$S_f(S_f^1, S_f^2)$	Surface where prescribed tractions are applied
S_c^1 and S_c^2	Potential contact surfaces of two contacting bodies
S_c	Common potential contact surface
S	Boundary of domain V
σ_k	Principal stress components, where $k = 1, 2$ in 2D and $k = 1, 2, 3$ in 3D
σ_{kj}	Stress tensor

$\sigma^*_{kj,j}$	Stress fundamental solutions
$[\sigma']$	Deviatoric stress tensor associated with $[\sigma]$
$\bar{t}_k$	Prescribed traction
t^*_k	Weighting (virtual) tractions
$t^*_{\ell k}$	k components of traction fundamental solutions
t^*_{kij}	Third order traction fundamental tensors
Tr	Transformation matrix
θ_c	Contact tangent angle
$(u^\alpha_{cn}, u^\alpha_{ct_1}, u^\alpha_{ct_2})$	Normal and tangential displacement components of two contacting bodies ($\alpha = 1, 2$)
u^r_{cn}	Elastic or reversible part of the normal relative displacement u_{cn}
u^{ir}_{cn}	Inelastic or irreversible part of the normal relative displacement u_{cn}
$u^r_{ct_\ell}$	Elastic or reversible (sticking) part of the tangential relative displacement u_{ct}
$u^{ir}_{ct_\ell}$	Inelastic or irreversible (sliding) part of the tangential relative displacement u_{ct}
u_0	Initial gap between the pairs of points on the potential contact surfaces of the two bodies before any loading application
$\bar{u}_k$	Prescribed displacement
$u_{k,j}$	Displacement derivatives
u^*_k	Weighting (virtual) displacements
$u^*_{\ell k}$	k components of displacement fundamental solutions
u^*_{kij}	Third order displacement fundamental tensors
$\{u_c\}$	Normal and tangential contact relative displacements
V^1 and V^2	Domains of bodies α ($\alpha = 1, 2$)
$V = V^1 \cup V^2$	Domain
χ	Coefficient to put $(\text{Yield})_{max}$ at last yield
Y	Uniaxial yield stress
$\{w\}$	Unknown vector
$\{z\}$	Unknown vector

Superscripts

α ($\alpha = 1, 2$)	Two linear elastic contacting bodies
c	Complementary
d	Dimension
e	Elastic
i	Location or interior
ir	Inelastic or irreversible
o	Inelastic
p	Plastic
r	Elastic or reversible
s	Surface (boundary)
T	Transpose
σ	Corresponding to the stresses

u	Corresponding to the displacements
$*$	Corresponding to the fundamental solutions
$'$	Deviatoric
$\bar{}$	Prescribed values

Subscripts

c	Contact
f	Corresponding to the applied tractions
i, j, k, ℓ	Indices defining the tensor components
k	Number of side in the polygonal projection of the polyhedral approximation of the friction cone
ℓ	Linearization
n	Normal
o	Outside
p	Plastic
r	Number of sides in the polygonal projection of the polyhedral approximation of the yield function
t	Tangential
u	Corresponding to the applied displacements
0	Initial
1, 2, 3	Number of components or bodies

Abbreviations

BEM	Boundary Element Method
BE-MP	Boundary Elements-Mathematical Programming
BIE	Boundary Integral Equation
FEM	Finite Element Method
LCP	Linear Complementarity Problem
MP	Mathematical Programming
PLCP	Parametric Linear Complementarity Problem
QP	Quadratic Program
RIP	Reduced-Integration Penalty Method
2D	Two dimensions
3D	Three dimensions

CHAPTER 1

Introduction

1.1 The contact problem

Contact problems are a common occurrence in engineering practice. Practical applications of such problems occur in the linear-elastic, small-displacement range, such as in the case of operating mechanical components, as well as in the essentially nonlinear, large-displacement, large-strain range, encountered in metal-forming operations.

Contact mechanics can be defined as the mechanics of the interaction of deformable bodies that touch each other in a contact zone, also referred to as a contact region or contact patch. The contact surface, with its accompanying load transfer, may well constitute the critical factor in the structural members. Thus it is essential to be able to perform contact stress analysis of a component accurately and efficiently. In other words, the analysis of contact problems is useful in improving the mechanical properties in the contact region and increasing the safety of contacted components.

Contact problems may be classified into three types depending on how the contact zone is configured. The first type of contact occurs when two contacting bodies are pressed together, contact is initially made at a point (or line in the case of plane problems), and as the applied load is increased the contact region grows (fig. 1.1a). Such contacts are called advancing and result when a punch with a curved bottom is pressed against the surface of another body. They possess the characteristic that the contact pressure falls continuously to zero towards the edges of the contact. In the case where the size of the contact does not change upon loading, or the boundaries of the contact

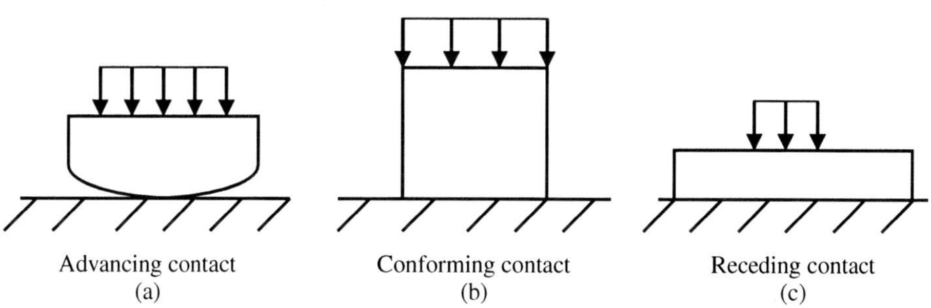

Figure 1.1: The various classes of contact.

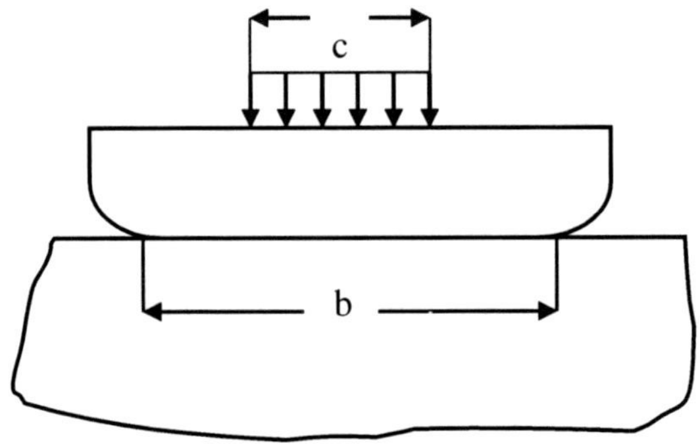

Figure 1.2: Contact problem with possibly receding, conforming or advancing contact.

surfaces in the loaded and unloaded configurations coincide, the contact is called conforming (fig. 1.1b). The contact involved in pressing a squat block against a half plane is obviously conforming. In this case the contact pressure distribution will be singular at each edge of the contact. Finally, if the contact dimension is decreased under an applied load, the contact is called receding (fig. 1.1c). For example when a thin strip is placed on top of an elastic half plane and loaded near its center, the ends of the strip separate from the half plane, resulting in a receding contact.

It is interesting to give an example (Dundurs and Stippes, 1970), in which there is a transition from receding through conforming to advancing contacts depending on various circumstances. As shown in fig. 1.2 a finite strip is resting on an elastic half plane. The bottom of the strip is flat near its center, but then curves away slightly from the half plane toward the ends. If the strip is loaded by uniformly distributed normal tractions over the length ‘c’ on its top surface, it is possible to distinguish two cases. First, when this length is small in comparison to the length ‘b’ of the straight part of the bottom profile, the strip can be expected to bend and pull away from that of the half plane under loading, and the contact is receding. In the second case if the distance ‘c’ is sufficiently large, the contact will be advancing.

The simplest contact problems occur when two smooth bodies with no friction come in contact with each other along a plane surface under static conditions. This is a linear problem as the contact area remains unchanged during loading and it is reversible as the system is conservative. When two curved bodies with smooth surfaces come into contact, the area of contact, in most cases, is not known *in advance* but also is a function of the load applied. With the change in contact area with load, the extent of contact is *a priori* unknown, rendering the problem nonlinear. However, it is still of a reversible nature due to the absence of nonconservative forces. When the contact area varies with load and when friction is also present, the system becomes nonlinear and irreversible. In these cases the contacts are elastic. In addition, other nonlinearities can arise from the material (elastoplastic behavior) and/or from the geometric model (large strains and deformations), resulting in plastic contacts. Such extensive mathematical models involve systems of inequalities or nonlinear equations, representing the most general case of an elastic and/or elastoplastic contact problem. Accurate stress and deformation analysis of such systems requires careful understanding of the frictional contact conditions as well as elastoplastic material behavior.

One should recognize that contact problems might also arise in the context of a single body, if one part of its surface can touch another part. Such self-contact is possible, for instance, in a slit hollow cylinder. There is, however, no essential difference between the contact of several bodies and self-contact.

1.2 A survey of analytical solutions to the contact problem

Heinrich Hertz (1896a,b) was the first to treat successfully the problem of the contact of smooth elastic bodies under normal loading in 1881. He computed and verified by experiment the load distributions over the contact area and solved for the stresses in the body through the use of the Newtonian potential function. Elastic contact problems are classified as Hertzian if they satisfy the following five conditions (Paul and Hashemi, 1980):

1. The bodies are homogeneous, isotropic, obey Hooke's law, and experience small strains and rotations (i.e., the linear theory of elasticity applies).
2. The contacting surfaces are frictionless.
3. The dimensions of the deformed contact patch remain small compared to the principal radii of the undeformed surfaces.
4. The deformations are related to the stresses in the contact zones as predicted by the linear theory of elasticity for half-planes or spaces.
5. The contacting surfaces are continuous, and may be represented by second-degree polynomials (quadratic surfaces) prior to deformation.

In practice, however, the Hertz theory is inadequate for treating the wide variety of technologically important problems. Three important restrictions apply. One is where the components are required to transmit tangential forces, as for example a locomotive wheel under driving or braking. The second is when the surfaces used are all irregular in relation to the scale of deflections. In this case, the surfaces cannot be modeled as locally quadratic. Problems of this type arise in roller bearings with crowned (i.e., rounded) edges, in ball bearings with small corrosion pits, in loosely fitted pin joints, etc. Thirdly, it is rare that the material is sufficiently hard to avoid some minimal plastic deformation. These types of contact problems are called non-Hertzian.

In a contact between two deformable bodies subjected to normal and subsequently to tangential loads, sliding will not start simultaneously over the whole contact area. For a given load, parts of the contact will slide and the rest will stick. This phenomenon is called micro-slip and is of significant importance in many applications where friction is involved, such as screw joints, couplings, shrink fits, rolling bearings, gears, power transmission, the railway industry, etc. It should be noted that a distinction exists between slip, denoting relative displacement of contiguous points on a portion of the contact surface, and sliding, which consists of some relative tangential displacement between the contacting bodies. In addition, repeated sliding contact may cause pitting, fretting, spalling or other fatigue failure. On the other hand, frictional contact problems constitute nonclassical (nonlinear) boundary value problems due to the frictional boundary conditions, which render them nonself-adjoint (Klarbring, 1986a, 1990). This type of contact problem has the following properties:

1. Frictional forces are inherently nonconservative.
2. Classical friction is a surface, non-smooth (non-associated) and path-dependent phenomenon, which is impossible to formulate as a purely static problem, even if inertia is neglected, and a discretization of the problem therefore produces nonsymmetrical matrices.
3. Solutions of quasi-static frictional contact problems are not necessarily unique.

Because of these properties and the nonlinear character of contact problems, due to the high traction that appears in the neighborhood of the contact zone, especially when this is small, the traditional Hertz theory no longer holds true, and the stress redistribution in the localized plastic region, which is produced not only by contact tractions but also by high traction gradients, requires a more rigorous approach, involving elastoplastic frictional contact analyses. It is obvious, therefore, that by including the plastic deformation in the formulation of frictional contact problems the calculation becomes even more complicated.

Both elastic and elastoplastic contact problems have long been of considerable interest. A much larger literature base exists covering a variety of elastic contact boundary value problems than elastoplastic cases. The presence of frictional forces and geometric nonlinearity places limits on the practical validity of the analytical study of elastic contact problems. Analytical solutions to frictionless contact problems with simple geometries are much easier to find than solutions to contact problems with friction. Excellent accounts of the classical analytical techniques used in the treatment of elastic contact problems, as well as the solutions to many interesting problems, can be found either in a wide range of books such as Galin (1953, 1980), Love (1944), Muskhelishvili (1963), Luré (1964), Duvaut and Lions (1972), Ling *et al.* (1973, 2002), de Pater and Kalker (1975), Rekach (1979), Gladwell (1980), Johnson (1985), Fabrikant (1989, 1991), Kalker (1990), Singer and Pollock (1992), Hills *et al.* (1993), Madhusudana (1996), Pfeiffer and Glocker (1996), Khludnev and Sokolowski (1997), Goryacheva (1998), Zambelli and Vincent (1998), Seireg (1998), Maugis (2000), Fischer-Cripps (2000, 2002), Alexandrov and Pozharskii (2002), and Sextro (2002) or in the papers of Thomas and Hoersch (1930), Kalker (1977) and Johnson (1982b), among others.

The literature on the analytical solution of tangential contact problems can be divided into two groups. The first contains publications on the contact between two stationary bodies, while the second group concerns publications on the contact between two moving bodies. In addition, for the contact between two moving bodies, one can distinguish between stationary motion where the stress field is independent of the time and transient motion where the stress field is time dependent. The first group has been investigated by, among others, Cattaneo (1938), Beeching and Nicholls (1948), M'Ewen (1949), Mindlin *et al.* (1949, 1951), Lubkin (1951), Smith *et al.* (1953), Johnson (1955), Deresiewicz (1957, 1958, 1961, 1974), Goodman (1962, 1982), Heems *et al.* (1963), Vermeulen and Johnson (1964), Spence (1973, 1975, 1986), Prasad and Dasgupta (1975), Keer *et al.* (1976, 1984), Raoof and Hobbs (1989), Mroz and Stupkiewicz (1992), Yan (1993), Jäger (1995, 1998, 1999), Bjorklund and Andersson (1994), Ciavarella (1998a,b,c), Faraji *et al.* (1999), and treated in the books of Kragelsky *et al.* (1982), Ludema (1996), and Hills and Nowell (1994). The second one has been considered by, among others, Carter (1926), Poritsky (1950), de Pater (1975), Kalker (1990) and treated in the books of Harris (1966), Jacobson and Kalker (2000), and Stolarski and Tobe (2001).

As seen from the literature, the frictional contact problem still remains one of the most difficult challenges in solid mechanics. In particular, 3D frictional contact problems have rarely been treated. Due to the difficulty in dealing analytically with frictional contact problems and particularly because no closed form solution is known for the elastoplastic case of the problem, most publications have resorted to numerical techniques.

1.3 A survey of computational and regularizational techniques to the contact problem

Many different formulations of contact problems with or without friction have been published based on the finite element and boundary element methods, which have both been used for numerical discretization purposes, over the last 20 years. Among the important procedures used are the following.

A. Stepwise incremental loading and iterative procedure

This approach uses finite element or boundary element concepts to evaluate the displacements of a known boundary value problem associated with a structure under frictional contact and differential loading. Boundary conditions at the (frictional) contact area are satisfied by iteration from the determined displacements. These displacements, caused by incremental loading, are stored and used to deform the structure to its current geometry. Successive incremental loading is applied until the total load (sum of the differential loads) is achieved, or until the structure fails according to some strength criterion. The method is simple and contains logical steps for establishing the contact area and the regions of slip and nonslip in the frictional case. In other words, implementation of this method need not require more sophisticated theories in applied mathematics and theoretical mechanics, when extending the finite (or boundary) element analysis to a linear elastic structure with the contact problem. Nevertheless, it possesses certain disadvantages such as the fact that the load distribution patches may significantly vary during the loading process, and the stress distributions that are critical for the structural design may occur at load levels that are lower in magnitude than the total applied load. In such situations, this method is inefficient for the determination of the deformation history of the bodies coming in contact experience during the process of load application.

B. Lagrange multiplier method

In this method, both displacement and contact pressure are regarded as independent variables, in which case the constraint and contact conditions can be strictly satisfied and the contact pressure obtained accurately when an appropriate mathematical formulation of the problem is used. Nevertheless, it has disadvantages including the fact that the number of unknowns varies with the change in the number of active contact nodes where contact actually takes place, and that the stiffness (associated tangent) matrix is indefinite and contains zero components along its diagonal. In order to overcome the difficulties of this method, either the perturbed Lagrangian multiplier method (which can be considered as a generalization of the Lagrange multiplier method where an additional term involving the contact tractions is added to the variational equations, and in addition, is equivalent to the penalty function method) or augmented Lagrange multiplier method is introduced.

C. Penalty function method

This method provides approximate solutions and the accuracy of the result is much influenced by the value of the penalty number. On the other hand, while a rather small penalty number usually leads to acceptable results, which may be achieved at the cost of slow convergence, numerical instability problems can arise if it is too large. This method has advantages such as a fixed number of unknowns, not requiring additional independent variables, easy implementation in existing analysis codes, no increase in the number of degrees of freedom in the final equations to be solved, a coefficient matrix of the system, which remains the same, and ready interpretation from the physical standpoint. Unfortunately, it is well known in this method that because the contact force is assumed to be proportional to the penetration at the nodal point, the approximating functions for the contact pressure are of the same order as the displacement functions. This may cause oscillations in the contact pressure distribution. A reduced-integration penalty method (RIP) (Kikuchi, 1982; Oden and Kikuchi, 1982) must then be used to eliminate such a numerical instability.

It should be noted that combination of the penalty approach and the Lagrangian multipliers method produces the augmented Lagrangian method, which has advantages over both methods by introducing no additional variables into the solution matrix, while providing better conditioning of the governing equations and exact satisfaction of the constraints using finite penalty numbers. Besides, it minimizes the disadvantages of the other two methods.

D. Flexibility matrix approach

This approach deals with the deformation analysis and the contact calculations separately. The size of the flexibility matrices, which represent the relationship between traction and displacement only at the nodes coming into contact, in conjunction with contact boundary conditions, can be computed separately for the contacting bodies. Therefore, the method depends only on the actual number of nodes in contact. The main advantage of the flexibility approach is that it is very easy and efficient to implement into boundary element codes. Once the flexibility matrix is computed it can be used throughout all the contact calculations if the deformation of bodies coming into contact is assumed to be small. Moreover, after substructuring, the elements of the flexibility matrix needed are those corresponding to the contacting nodes only. This results in a relatively small size flexibility matrix and consequently in considerable saving in computing time. The main disadvantage of this method is that it applies only locally, in the contact zone.

In order to solve the contact problems, these procedures, in general, may employ one of two algorithms, namely: iterative (trial and error) methods and mathematical programming problems such as the direction method (Kinderlehrer and Stampacchia, 1980; Panagiotopoulos, 1985). Iterative methods firstly consider three possible states at a point of the potential contact area, which are: separation (there is no contact between the bodies); stick (there is contact but no relative tangential motion); and slip state (there is contact and relative tangential motion). Secondly, they analyze the problem assuming one of the three states at each contact point, then solve the resulting problem and check the assumptions afterwards. If they turn out to be wrong, new assumptions have to be made according to some prescribed rules, a new problem is solved and a new iteration obtained. Numerical and mathematical studies have shown that these methods are not only time consuming and do not generally guarantee convergence to the correct solution (Stavroulakis *et al.*, 1991) but also are costly procedures for determining the contact area, especially for sliding contact with friction (Chen and Yen, 1988). For these reasons, algorithms of the second category, which have the property of stable convergence, have been applied in structural mechanics for the analysis of problems with inequalities such as contact problems. They are accurate and quick. To give a measure of computational economy for 1000 contact unknowns, the computational time of mathematical programming versus trial and error methods is 1/20–1/25 (Stavroulakis *et al.*, 1991). They are based on inequality constrained minimum propositions for the potential and/or the complementarity energy, which results from the variational formulation of the problem. In particular, they reformulate the contact problem as a mathematical problem, or as a problem which is in a sense close to such a problem. Usually such well-known mathematical problems may be found in the field of mathematical programming (MP), with particular structures involving linear complementarity problems (LCP), parametric linear complementarity problems (PLCP), quadratic programs (QP), nonlinear programming problems or variational inequalities. It should be noted that a difficulty involving frictional contact problems using nonlinear programming theories is that the virtual work of the frictional forces is non-differentiable and nonlinear, and the uniqueness of the solution of the variational inequality for the general contact with Coulomb friction has not yet been proven (Sha *et al.*, 1992).

1.4 A survey of numerical discretization methods applied to the contact problem

Along with the development of computational mechanics technology, the finite element method (FEM) (also known as the domain method) represents the most widely used numerical tool, especially for the numerical treatment of contact problems. In this approach, the domain has to be divided into small but finite elements, in which variables are interpolated by polynomials. Using suitable variational principles, the minimum energy distribution is obtained. This method also produces an algebraic system of equations, having narrow band matrices.

Up to now, based on the FEM, most publications have concentrated on elastic contact problems. References include, among others, the books of Kikuchi and Oden (1988), the special issue of the *Journal de Mécanique Théorique et Appliquée* edited by Raous *et al.* (1988) and Zhong (1993), and papers by Chan and Tuba (1971), Francavilla and Zienkiewicz (1975), Fredricksson (1976), Okamoto and Nakazawa (1979), Sachdeva and Ramakrishnan (1981), Oden and Pires (1983), Oden and Carey (1984), Bathe and Chaudhary (1985), Chaudhary and Bathe (1986), Klarbring (1986a to 1993), Klarbring and Bjorkman (1988, 1992), Lee (1989), Bjorkman *et al.* (1995), Heege and Alart (1996). As far as elastoplastic contact problems are concerned, publications include, among others, Hardy *et al.* (1971), Torstenfelt (1983, 1984), Valliappan *et al.* (1984), Lee and Kwak (1984), Follansbee and Sinclair (1984), Cheng and Kikuchi (1985), Sinclair *et al.* (1985), Endahl (1986), Joo and Kwak (1986), Tsai and Chen (1986), Zhong and Sun (1988), Alart and Curnier (1991, 1992), Saran and Taylor (1992), Zhong and Wu (1993), Zhu (1995a), and Vu-Quoc and Zhang (1999). In addition, the books of Crisfield (1997, 2000), Belytschko *et al.* (2000), Laursen (2002), Sofonea and Han (2002), Wriggers (2002), Mohammadi (2003), Jäger(2004), and the special issue of the *International Journal of Applied Mathematics and Computer Science* edited by Sofonea and Viano (2002) can be referred to.

It is interesting to note that multipurpose finite element codes exist such as ABAQUS (Hibbitt *et al.*, 2001), which use an elastic–plastic constitutive model based on the assumption that the deformation rate can be decomposed into elastic and plastic parts by a multiplicative decomposition. In order to justify such decomposition, it is necessary to ensure that a stress-free configuration can be obtained without additional plastic deformation by the inverse of the elastic part of the deformation, which in turn requires that the stress at each point remain inside the yield surface.

Although the FEM is very popular for engineering applications it also presents drawbacks. The major drawbacks of the FEM to derive the solution of contact problems are as follows:

1. Modeling entirely both contacting bodies with a sufficiently fine mesh discretization of the domains to ensure accurate results (leading to an uneconomical, if not intractable, analysis).
2. Having a high number of variables, high data preparation times, large coefficient matrix and long execution times.
3. Inserting special contact elements between the contacting bodies to model them.
4. Providing little motivation to model both contacting bodies in their entirety to determine the extent of the contact area and the contact stresses.
5. Inability of the classical FE displacement-based equations to satisfy simultaneously both equilibrium and compatibility conditions at the contact interface.

In recent years, the boundary integral equation (BIE) or boundary element method (BEM) has become available as an alternative numerical procedure for a number of nonlinear as well as classical linear boundary value problems. In this respect, the BEM is ideally suited to solve contact problems, since contact occurs only on boundaries, which are the primary concern of the boundary element solution procedure. Elastic contact solutions based on BEM have long been of interest and a large body of literature is available. The books of Karami (1989), Takahashi (1991), Antes and Panagiotopoulos (1992), Becker (1992), Man (1994), Aleynikov (2000) and the series of proceeding of the *International Conference on Computational Methods in Contact Mechanics I–VI* edited by Aliabadi and Brebbia (1993), Aliabadi and Alessandri (1995), Samartin and Aliabadi (1997), Gaul and Brebbia (1999), Dominguez and Brebbia (2001), Brebbia (2003), and papers by Andersson *et al.* (1980, 1981, 1982), Abdul-Mihsein *et al.* (1986), Kuich (1986), Kwak and Lee (1988, 1991), Yamazaki *et al.* (1988, 1993, 1994), Man *et al.* (1991), Dandekar and Conant (1992a,b), Blazques *et al.* (1992), Simunovic and Saigal (1992, 1995a,b), Kong *et al.* (1992), Gakwaya *et al.* (1992), Olukoko *et al.* (1993), Foces *et al.* (1993), Paris *et al.* (1995), Zhu (1995b), Xu and Kong (1996) can be referred to. The elastoplastic contact problem has attracted less attention. Some available references are: Tsuta and Yamaji (1983), Alcantud *et al.* (1990), Karami (1991), Kong *et al.* (1993), Martin *et al.* (1994), Huesmann and Kuhn (1995), Polizzotto and Zito (1998), and Gun (1999).

As seen from the literature, the BEM is well suited to the analysis of contact problems, offering the following advantages over current numerical methods such as the FEM:

1. Permitting modeling of only the boundaries of the contacting bodies and in the case of local plasticity the region where plastic deformation is expected, allowing a fine discretization of the contact zone and plastic region whilst still providing accurate, economical results and generally producing more accurate results since the approximations are confined to the boundary.
2. Reducing considerably data preparation time, number of equations and size of the coefficient matrix and solution time.
3. Imposing the exact contact constraints along the interface without the introduction of additional variables because of writing explicitly the contact equations, which contain both tractions and displacements as unknowns, and in addition computing both with the same degree of accuracy.
4. Calculating directly surface tractions in the contact zone, which is the most important part of the analysis.
5. Ability to efficiently capture the traction singularities that may occur on contact boundaries.
6. Possibility of coupling the normal and tangential tractions in the matrix system.

However, for elastoplastic contact analysis, material nonlinearities may require a domain discretization but, using the particular integrals formulation (meshless method) introduced by Henry and Banerjee (1988) or the application of dual reciprocity method (meshless method) represented by Nardini and Brebbia (1982), one can eliminate the need for volume integrals. Therefore, the major significant advantage of the BEM over the FEM is that only a surface discretization is maintained.

CHAPTER 2

Three-dimensional elastic frictional contact analysis, Part 1: Theory

2.1 Introduction

A general elastic analysis method is presented for 2- and 3D contact problems between two deformable bodies undergoing small displacements with or without friction (Faraji *et al.*, 2003a). In accordance with the analogy with plasticity, the contact boundary conditions are expressed as a linear complementarity form with penalty factors. For numerical discretization purposes, the boundary element method is employed. Based on the decoupling of normal pressure and tangential traction, the contact problem is transformed into a linear complementarity problem (LCP) that is solved by Lemke's algorithm. This method can avoid the tedious and costly classical iterative procedures and, if there is no unloading, only one incremental step is required for a solution. Finally, the analysis is successfully implemented through a package called CONTACTA. The dimension of the contact patch, the normal and tangential tractions distributed in the contact zone as well as the stresses and displacements for each node of contacting bodies can be evaluated using this package. One may consider this work as an LCP for an elastic frictional contact problem, employing an elastic perfectly plastic type of friction law rather than the more complex parametric LCP obtained by Gakwaya *et al.* (1992) and Kong *et al.* (1992), which were based on a rigid plastic model of friction law. It should be noted that a similar boundary element method (BEM) model of this analysis in two dimensions has already been presented by Zhu (1995b).

2.2 Contact boundary conditions between two bodies

Let us first consider the physical modeling of the contact problem between two linear elastic bodies α ($\alpha = 1, 2$), occupying an open set V of space R^d ($d = 2, 3$), as indicated in fig. 2.1(a).

Upon the application of loads (tractions or displacements), these bodies are brought into contact at least at one point. The domains of bodies α ($\alpha = 1, 2$) are V^1 and V^2 respectively, with $V = V^1 \cup V^2$. The boundary S of V may be decomposed into disjoint parts $S_u(S_u^1, S_u^2)$, $S_f(S_f^1, S_f^2)$, S_c^1 and S_c^2. Furthermore, one can define a common potential contact surface S_c located at the intersection of S_c^1 and S_c^2 and establish the following relations:

$$S = S_u \cup S_f \cup S_c, \qquad S_u \cap S_f = S_f \cap S_c = S_u \cap S_c = \emptyset. \tag{2.1}$$

Prescribed displacements and tractions are applied on S_u and S_f respectively, and the surfaces S_c^1 and S_c^2 are potential contact surfaces, which are chosen on the premises that:

1. The distance between S_c^1 and S_c^2 is small and of the order of the displacement.
2. Assuming a point-to-point approach, for each point on S_c^1, there exists a point on S_c^2 such that their outward normal vectors are almost parallel.

Then, in order to derive the contact boundary conditions between two bodies, we introduce a local coordinate system (ξ, η, ζ) with unit basis vectors $(\mathbf{t}_1, \mathbf{t}_2, \mathbf{n})$ on the potential contact surface $S_c(S_c = S_c^1 = S_c^2)$, where $\mathbf{n}$ defines the normal vector (to body 1) and $\mathbf{t}_1$ and $\mathbf{t}_2$ the two tangential directions (see fig. 2.1(a,b)). Based on this modeling, the following conditions for the pairs of contact points on the potential contact surface are made.

A. Unit normal and tangential vectors

Following the second assumption above, the contact surfaces S_c^α have approximately common outward unit normals (fig. 2.1b) so we can write:

(a) In 2D:

$$\begin{Bmatrix} \mathbf{n} \\ \mathbf{t} \end{Bmatrix} = \begin{Bmatrix} \mathbf{n}^1 \\ \mathbf{t}^1 \end{Bmatrix} = -\begin{Bmatrix} \mathbf{n}^2 \\ \mathbf{t}^2 \end{Bmatrix}. \tag{2.2a}$$

(b) In 3D:

$$\begin{Bmatrix} \mathbf{n} \\ \mathbf{t}_1 \\ \mathbf{t}_2 \end{Bmatrix} = \begin{Bmatrix} \mathbf{n}^1 \\ \mathbf{t}_1^1 \\ \mathbf{t}_2^1 \end{Bmatrix} = -\begin{Bmatrix} \mathbf{n}^2 \\ \mathbf{t}_1^2 \\ \mathbf{t}_2^2 \end{Bmatrix}. \tag{2.2b}$$

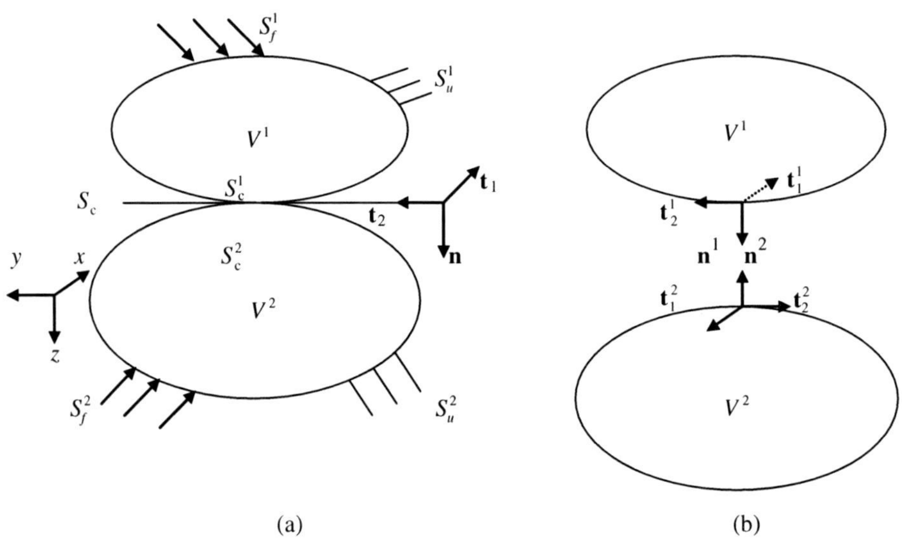

Figure 2.1: Two bodies in contact.

B. Normal and tangential relative displacement components

Contact phenomena take place in the deformed state that is the final stage of deformation from the undeformed state. The undeformed and deformed states of the two bodies α ($\alpha = 1, 2$) are shown in fig. 2.2 by the full and broken horizontal lines. Referring to this figure and following the small displacement and strain theory, and the introduction of the displacement vectors on the potential contact surfaces of the two bodies, one can express the normal and tangential relative displacement components on S_c as follows:

(a) In 2D:

$$\begin{Bmatrix} u_{cn} \\ u_{ct} \end{Bmatrix} = \begin{Bmatrix} u^2_{cn} - u^1_{cn} + u_0 \\ u^2_{ct} - u^1_{ct} \end{Bmatrix}. \tag{2.3a}$$

(b) In 3D:

$$\begin{Bmatrix} u_{cn} \\ u_{ct_1} \\ u_{ct_2} \end{Bmatrix} = \begin{Bmatrix} u^2_{cn} - u^1_{cn} + u_0 \\ u^2_{ct_1} - u^1_{ct_1} \\ u^2_{ct_2} - u^1_{ct_2} \end{Bmatrix}, \tag{2.3b}$$

where (u^α_{cn}, $u^\alpha_{ct_1}$, $u^\alpha_{ct_2}$) are the normal and tangential displacement components of both bodies and u_0 is the normal initial gap between the pairs of points on the potential contact surfaces of the two bodies before any loading application.

Following the first relation of eqn (2.3), where $u_{cn} = u^2_{cn} - u^1_{cn} + u_0$, three cases can be distinguished:

1. $u_{cn} > 0$: no-penetration between the two bodies; (2.4a)
2. $u_{cn} = 0$: contact between the two bodies; (2.4b)
3. $u_{cn} < 0$: penetration between the two bodies. (2.4c)

The third case can occur when welding takes place between two bodies, but this penetration deviates from ordinary mechanical contact theory and practice and we do not consider penetration phenomena to be a part of contact mechanics. This leads us to the no-penetration condition that is the so-called first idealized Signorini–Fichera condition (1933, 1959, 1963, 1964) for normal contact:

$$u_{cn} \geq 0. \tag{2.5}$$

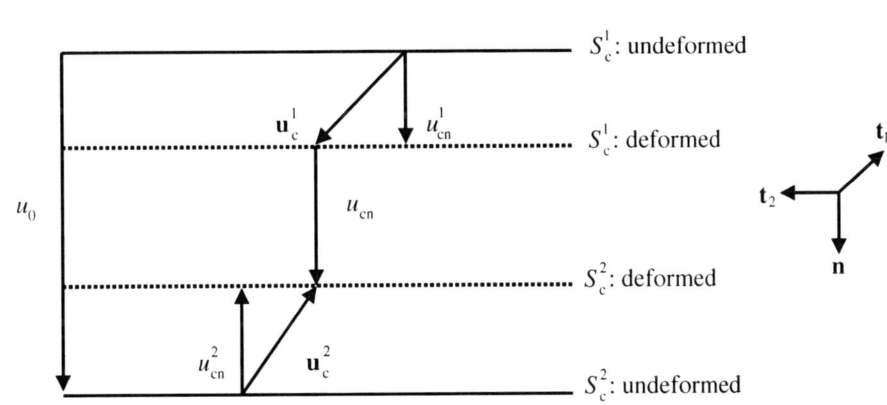

Figure 2.2: Undeformed and deformed state of two contacting bodies.

C. Normal and tangential contact traction components

As with to the relative displacement vectors, we can define the contact traction vectors on the potential contact surfaces of both bodies from the following relations:

(a) In 2D:

$$\begin{Bmatrix} p_{\text{cn}} \\ p_{\text{ct}} \end{Bmatrix} = -\begin{Bmatrix} p^1_{\text{cn}} \\ p^1_{\text{ct}} \end{Bmatrix} = \begin{Bmatrix} p^2_{\text{cn}} \\ p^2_{\text{ct}} \end{Bmatrix}. \tag{2.6a}$$

(b) In 3D:

$$\begin{Bmatrix} p_{\text{cn}} \\ p_{\text{ct}_1} \\ p_{\text{ct}_2} \end{Bmatrix} = -\begin{Bmatrix} p^1_{\text{cn}} \\ p^1_{\text{ct}_1} \\ p^1_{\text{ct}_2} \end{Bmatrix} = \begin{Bmatrix} p^2_{\text{cn}} \\ p^2_{\text{ct}_1} \\ p^2_{\text{ct}_2} \end{Bmatrix}. \tag{2.6b}$$

The normal contact traction (pressure) must satisfy the compressive condition, which is the second Signorini–Fichera condition for normal contact, written:

$$p_{\text{cn}} \geq 0 \qquad \text{(compression taken as positive).} \tag{2.7}$$

Note that the impenetrability relation (eqn (2.5)) and the no-tension condition (eqn (2.7)) may be combined in the following orthogonality property or complementarity condition that is the third Signorini–Fichera condition:

$$p_{\text{cn}} u_{\text{cn}} = 0. \tag{2.8}$$

D. Friction law

The basic principles of friction may be found in Leonardo da Vinci's notes, dating from the 15th century. The next to study friction was Amontons in the 17th century (1699). However, the theory of friction is commonly dated from Coulomb's paper (1785). In what follows, the isotropic Coulomb's friction law will be assumed. This law states that slipping takes place if the tangential traction is equal to a certain threshold value:

$$|p_{\text{ct}}| \leq \mu\, |p_{\text{cn}}|\,, \tag{2.9}$$

where μ is the coefficient of friction ($\mu \geq 0$). The graphical representation of Coulomb's friction law in 2D and 3D, which is called the slip function or friction criterion, is shown in fig. 2.3. This function is represented by two straight lines in 2D and by a conical surface in 3D whose boundary is the circle of radius (μp_{cn}). The contact tractions corresponding to the sticking condition lie between the two limit lines in 2D and the interior of the cone in 3D, and for the sliding condition on the boundary lines in 2D and on the conical surface in 3D. Otherwise, the tangential traction state is not admissible in Coulomb's model.

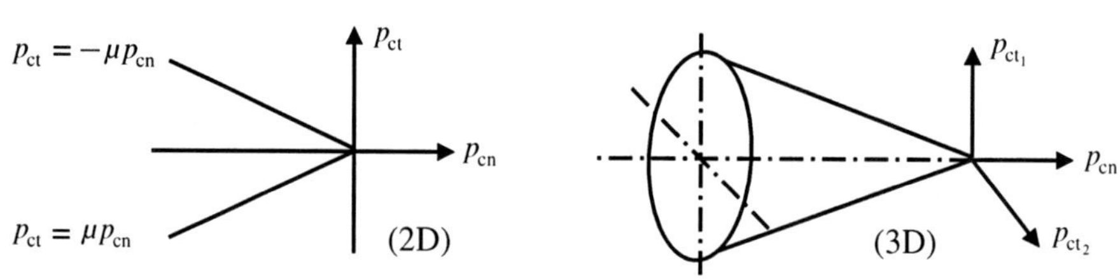

Figure 2.3: Graphical representation of Coulomb's friction law in 2D and 3D.

2.3 Elastoplastic analogy for a model of contact boundary conditions

The criterion for contact may be expressed in the normal and tangential direction:

(1) *Normal direction.* Relation between u_{cn} and p_{cn}:

$$\begin{cases} \text{if } p_{cn} = 0, & \text{then } u_{cn} \geq 0 \quad \text{(no contact condition)}, \\ \text{if } p_{cn} \geq 0, & \text{then } u_{cn} = 0 \quad \text{(contact condition)}. \end{cases} \tag{2.10}$$

This relation, in addition to eqn (2.8), expresses the normal unilateral contact law and it is shown in fig. 2.4. The law is characterized by a geometric condition of no-penetration, a static condition of no-tension and a mechanical complementarity condition previously classified as the Signorini–Fichera condition. This possible normal contact state is similar to the behavior of an ideal rigid plastic material. Using this concept leads us to the normal separation (yield) function in the positive normal direction for such contact problems:

$$f_{c1} = -p_{cn} \leq 0. \tag{2.11}$$

(2) *Tangential direction.* Relation between u_{ct} and p_{ct}:

$$\begin{cases} \text{if } |p_{ct}| \leq \mu\,|p_{cn}|, & \text{then } u_{ct} = 0 \quad \text{(sticking contact condition)}, \\ \text{if } |p_{ct}| = \mu\,|p_{cn}|, & \text{then } |u_{ct}| \geq 0 \quad \text{(sliding contact condition)} \end{cases} \tag{2.12a}$$

and in addition, the sliding contact condition can be expressed as:

$$\begin{cases} \text{if } p_{ct} = \mu p_{cn}, & \text{then } u_{ct} \geq 0 \quad \text{(positive slip condition)}, \\ \text{if } p_{ct} = -\mu p_{cn}, & \text{then } u_{ct} \leq 0 \quad \text{(negative slip condition)}. \end{cases} \tag{2.12b}$$

In fact, these relations have to be written in terms of velocities, instead of the relative tangential displacements, because the friction phenomena clearly depend on the whole loading history. Nevertheless, the static Coulomb problem considered here will satisfy the solution when the loadings are monotonic and proportional, as proposed in the present study. It is interesting to stress that most of the mathematical results so far established have been based on this static formulation. One may add to the static formulation (eqns (2.12a,b)), the following complementarity condition:

$$|u_{ct}|\,(|p_{ct}| - \mu p_{cn}) = 0 \tag{2.13}$$

and, using these relations, the sliding threshold contact law can be represented as in fig. 2.5.

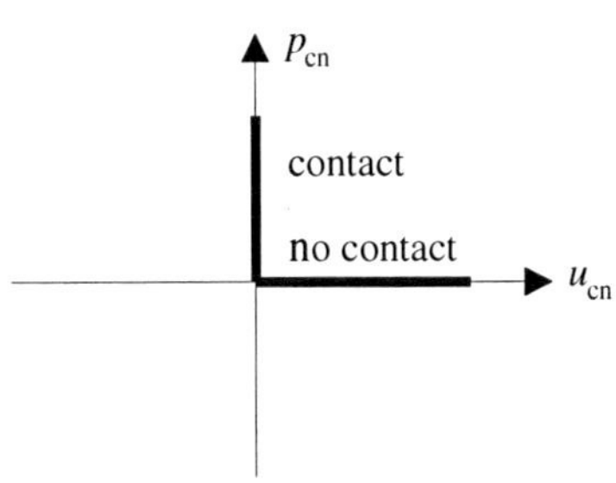

Figure 2.4: Normal unilateral contact law.

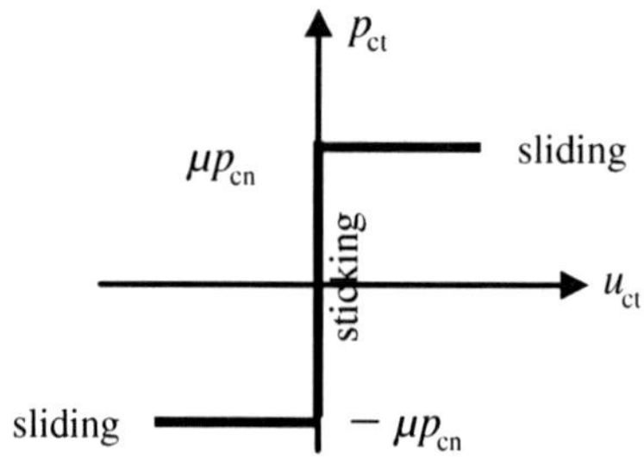

Figure 2.5: Sliding threshold contact law.

The law is characterized by a static friction criterion, a sliding threshold rule and a mechanical complementarity condition. This possible tangential contact state is similar to the behavior of an ideal rigid plastic material. It leads us to the tangential slip (yield) functions in the horizontal–tangential directions for contact problems:

$$\{f_c\} = \{|p_{ct}| - \mu p_{cn}\} \leq 0. \tag{2.14}$$

This relation, in fact, is the Coulomb's classical law of friction that may be approximated in 3D by a piecewise linear formulation (an n-side polytope law), which is required for numerical solution. A linearization of this law, as shown in fig. 2.6, in which an eight-sided discretization of the friction cone (eight-sided friction polytope) is adopted, is described by the following formulations (Klarbring, 1986a,b; Gakwaya *et al*., 1992):

$$f_{c(k+1)} = [\mu \quad \cos\beta_k \quad \sin\beta_k] \begin{Bmatrix} -p_{cn} \\ p_{ct_1} \\ p_{ct_2} \end{Bmatrix}, \tag{2.15a}$$

where

$$\beta_k = (k-1)\frac{2\pi}{m_c}, \quad k = 1, \ldots, m_c, \quad m_c = 2 \text{ (in 2D)}, \quad m_c = 8 \text{ (in 3D)}. \tag{2.15b}$$

Geometrically, β_k represents the direction of the normal vector to side k in the polygonal projection, as shown in fig. 2.6. Notice that other polyhedral approximations of the friction cone can also be used.

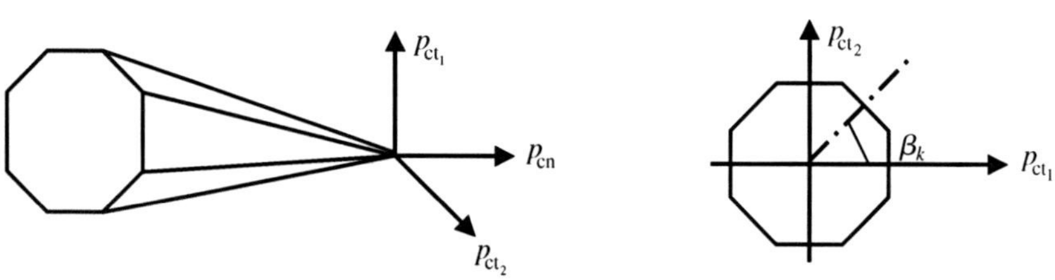

Figure 2.6: A linearization form of Coulomb's friction law in 3D.

We now express the normal and tangential contact separation-slip (yield) functions together as an explicit matrix form using eqns (2.11, 2.15a,b):

(a) In 2D:

$$\begin{Bmatrix} f_{c1} \\ f_{c2} \\ f_{c3} \end{Bmatrix} = \begin{Bmatrix} -p_{cn} \\ p_{ct} - \mu p_{cn} \\ -p_{ct} - \mu p_{cn} \end{Bmatrix} = \begin{bmatrix} 1 & 0 \\ \mu & \cos 0 \\ \mu & \cos \pi \end{bmatrix} \begin{Bmatrix} -p_{cn} \\ p_{ct} \end{Bmatrix}. \tag{2.16a}$$

(b) In 3D:

$$\begin{Bmatrix} f_{c1} \\ f_{c2} \\ f_{c3} \\ f_{c4} \\ f_{c5} \\ f_{c6} \\ f_{c7} \\ f_{c8} \\ f_{c9} \end{Bmatrix} = \begin{bmatrix} 1 & 0 & 0 \\ \mu & \cos 0 & \sin 0 \\ \mu & \cos \pi/4 & \sin \pi/4 \\ \mu & \cos \pi/2 & \sin \pi/2 \\ \mu & \cos 3\pi/4 & \sin 3\pi/4 \\ \mu & \cos \pi & \sin \pi \\ \mu & \cos 5\pi/4 & \sin 5\pi/4 \\ \mu & \cos 6\pi/4 & \sin 6\pi/4 \\ \mu & \cos 7\pi/4 & \sin 7\pi/4 \end{bmatrix} \begin{Bmatrix} -p_{cn} \\ p_{ct_1} \\ p_{ct_2} \end{Bmatrix}. \tag{2.16b}$$

Finally, we can conclude that using the theoretical framework of material plasticity, the contact separation-slip (yield) functions have been derived and are rearranged below in a unified matrix form:

$$\{f_c\} = [C_\ell]\{p_c\}, \tag{2.17a}$$

where $\{f_c\}$ are contact separation-slip (yield) functions, $[C_\ell]$ the linearization matrix and $\{p_c\}$ the normal and tangential contact tractions, and, in which:

(a) In 2D:

$$\{f_c\} = \begin{Bmatrix} f_{c1} \\ f_{c(k+1)} \end{Bmatrix},$$

$$[C_\ell] = \begin{bmatrix} 1 & 0 \\ [\mu & \cos \beta_k] \end{bmatrix}, \quad k \text{ and } \beta_k \text{ are given by eqn (2.15b)}, \tag{2.17b}$$

$$\{p_c\} = \begin{Bmatrix} -p_{cn} \\ p_{ct} \end{Bmatrix}.$$

(b) In 3D:

$$\{f_c\} = \begin{Bmatrix} f_{c1} \\ f_{c(k+1)} \end{Bmatrix},$$

$$[C_\ell] = \begin{bmatrix} 1 & 0 & 0 \\ [\mu & \cos \beta_k & \sin \beta_k] \end{bmatrix}, \quad k \text{ and } \beta_k \text{ are given by eqn (2.15b)}, \tag{2.17c}$$

$$\{p_c\} = \begin{Bmatrix} -p_{cn} \\ p_{ct_1} \\ p_{ct_2} \end{Bmatrix}.$$

2.4 Penalty complementarity contact boundary conditions

We chose the penalty formulation as the most simple and easiest procedure to be implemented in a finite element as well as in a boundary element program. Despite the well-known problems inherent in the selection of suitable penalty numbers, the technique has been found to be most convenient for the solution of boundary value problems subject to constraints, and is therefore widely used for contact problems. In consequence, using the penalty formulation, following concepts drawn from the theory of plasticity and applying the contact separation-slip (yield) functions given in Section 2.3, the penalty complementarity contact conditions will be presented.

The penalty formulations, for normal unilateral as well as for sliding threshold contact law (Curnier, 1984, 1999; Oden and Martins, 1985; Klarbring, 1986b, 1999; Raous, 1999), can be expressed as follows.

A. Penalty formulation for normal unilateral contact law (Curnier, 1984, 1999)

By replacing the no-penetration condition shown in fig. 2.4 by the penalty penetration illustrated in fig. 2.7, a penalty formulation between the normal contact traction p_{cn} and the normal relative displacement u_{cn} may be derived. By means of fig. 2.7 and similar to the elastoplastic material behavior, the normal relative displacement u_{cn} may be divided into an elastic or reversible part $u_{\mathrm{cn}}^{\mathrm{r}}$ and an inelastic or irreversible part $u_{\mathrm{cn}}^{\mathrm{ir}}$:

$$u_{\mathrm{cn}} = u_{\mathrm{cn}}^{\mathrm{r}} + u_{\mathrm{cn}}^{\mathrm{ir}}, \tag{2.18}$$

where $u_{\mathrm{cn}}^{\mathrm{r}}$ is related to p_{cn} by introducing a normal penalty parameter $E_{\mathrm{n}}(E_{\mathrm{n}} \rightarrow \infty)$:

$$u_{\mathrm{cn}}^{\mathrm{r}} = -p_{\mathrm{cn}}/E_{\mathrm{n}}, \tag{2.19}$$

in which $E_{\mathrm{n}}(E_{\mathrm{n}} \rightarrow \infty)$ is illustrated in fig. 2.7 and physically penalizes the gap in the normal direction.

$u_{\mathrm{cn}}^{\mathrm{ir}}$ can be determined using the associated flow rule for normal contact:

$$u_{\mathrm{cn}}^{\mathrm{ir}} = \lambda_{\mathrm{c1}} \frac{\partial g_{\mathrm{c1}}}{\partial p_{\mathrm{cn}}}, \tag{2.20}$$

where $g_{\mathrm{c1}} = -p_{\mathrm{cn}}$ is the flow potential function for normal contact corresponding to the chosen separation (yield) function (eqn (2.11)). λ_{c1} is called the flow parameter for normal contact,

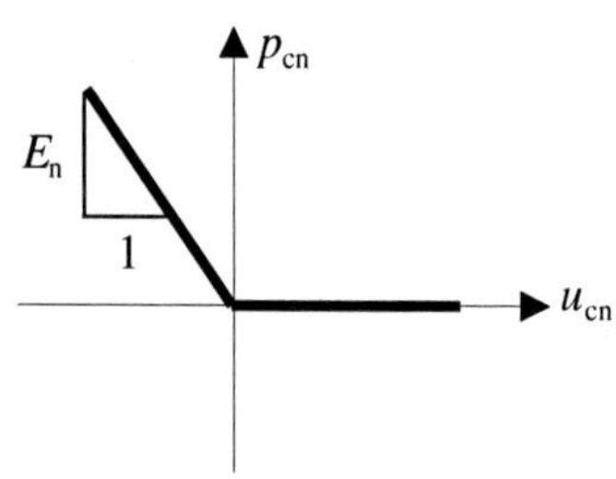

Figure 2.7: Normal unilateral penalty contact law.

or normal contact multiplier, which denotes the separation distance in normal direction $\mathbf{n}$ that is a non-negative scalar to be determined. It should be noted that the following complementarity relation exists between the contact flow parameter λ_{c1} and the first separation (yield) function f_{c1}:

$$\lambda_{c1} \begin{cases} = 0, & \text{when } f_{c1} \leq 0, \\ \geq 0, & \text{when } f_{c1} = 0. \end{cases} \tag{2.21}$$

Finally, using eqns (2.18–2.21), the following penalty formulation can be derived between the normal contact traction p_{cn} and the normal relative displacement u_{cn}:

$$-p_{cn} = E_n(u_{cn} + \lambda_{c1}). \tag{2.22}$$

It is seen that, in the limit case as ($E_n \rightarrow \infty$), the no-penetration condition (fig. 2.4) is satisfied.

B. Penalty formulation for sliding threshold contact law (Raous, 1999)

Similarly, we can also derive a penalty formulation between tangential contact traction p_{ct} and tangential relative displacement u_{ct} by replacing the classical Coulomb's friction law (fig. 2.5), by its penalty regularization shown in fig. 2.8. From fig. 2.8 and similar to the elastoplastic material behavior, the tangential relative displacement u_{ct} may be additively decomposed into an elastic or reversible (sticking) part $u^{r}_{ct\ell}$ and an inelastic or irreversible (sliding) part $u^{ir}_{ct\ell}$:

$$u_{ct\ell} = u^{r}_{ct\ell} + u^{ir}_{ct\ell}, \quad \text{where} \begin{cases} \text{in 2D:} & \text{without } \ell, \\ \text{in 3D:} & \ell = 1, 2 \end{cases} \tag{2.23}$$

in which $u^{r}_{ct\ell}$ is related to $p_{ct\ell}$ by introducing a tangential penalty parameter $E_t(E_t \rightarrow \infty)$ such that:

$$u^{r}_{ct\ell} = p_{ct\ell}/E_t, \tag{2.24}$$

where $E_t(E_t \rightarrow \infty)$ is illustrated in fig. 2.8 and physically penalizes the slip in the tangential directions.

$u^{ir}_{ct\ell}$ can be expressed by a non-associated flow rule for tangential contact:

$$u^{ir}_{ct\ell} = \sum_{i=2}^{k+1} \lambda_{ci} \frac{\partial g_{ci}}{\partial p_{ct\ell}}, \tag{2.25a}$$

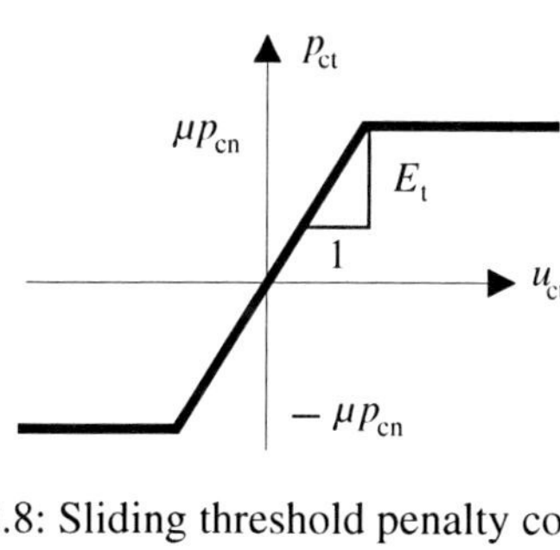

Figure 2.8: Sliding threshold penalty contact law.

where

(a) In 2D:

$$\text{without } \ell: \quad k=2; \;\; g_{\mathrm{ci}} = p_{\mathrm{ct}} \cos\beta_{i-1}, \;\; \beta_{i-1} \text{ is given by eqn (2.15b).} \tag{2.25b}$$

(b) In 3D:

$$\ell = 1,2: \quad k=8; \;\; g_{\mathrm{ci}} = p_{\mathrm{ct}_1} \cos\beta_{i-1} + p_{\mathrm{ct}_2} \sin\beta_{i-1}, \;\; \beta_{i-1} \text{ is given by eqn (2.15b).} \tag{2.25c}$$

The flow potential functions g_{ci} for tangential contact as chosen are different from the slip (yield) functions shown in eqn (2.15a), so we have what are called non-associated flow rules for the tangential contact. $\lambda_{\mathrm{c}(k+1)}$ are called the tangential flow parameters or tangential contact multipliers and represent the slip displacement(s) in the tangential direction(s). These are non-negative scalar parameters to be determined. It should be noted that the following complementarity relation exists between the contact consistency parameter, $\lambda_{\mathrm{c}(k+1)}$, and the slip (yield) functions, $f_{\mathrm{c}(k+1)}$:

$$\lambda_{\mathrm{c}(k+1)} \begin{cases} =0, & \text{when } f_{\mathrm{c}(k+1)} \le 0, \\ \ge 0, & \text{when } f_{\mathrm{c}(k+1)} = 0, \end{cases} \qquad \text{where } k = 1{:}2 \text{ in 2D and } k = 1{:}8 \text{ in 3D.} \tag{2.26}$$

Finally, using eqns (2.23–2.26) one can obtain the following penalty formulation between tangential contact traction p_{ct} and tangential relative displacement u_{ct}:

(a) In 2D:

$$p_{\mathrm{ct}} = E_{\mathrm{t}}(u_{\mathrm{ct}} - \lambda_{\mathrm{c}2} + \lambda_{\mathrm{c}3}). \tag{2.27a}$$

(b) In 3D:

$$\begin{Bmatrix} p_{\mathrm{ct}_1} \\ p_{\mathrm{ct}_2} \end{Bmatrix} = \begin{bmatrix} E_{\mathrm{t}} & 0 \\ 0 & E_{\mathrm{t}} \end{bmatrix} \left\{ \begin{Bmatrix} u_{\mathrm{ct}_1} \\ u_{\mathrm{ct}_2} \end{Bmatrix} - \sum_{k=2}^{9} \begin{bmatrix} \cos\beta_{k-1} \\ \sin\beta_{k-1} \end{bmatrix} \{\lambda_{\mathrm{c}k}\} \right\}. \tag{2.27b}$$

If $E_{\mathrm{t}} \to \infty$ the Coulomb's classical friction law (fig. 2.5) can be satisfied.

Now, an explicit matrix form of the penalty formulation for frictional contact problems may be written by assembling eqns (2.22, 2.27a,b):

(a) In 2D:

$$\begin{Bmatrix} -p_{\mathrm{cn}} \\ p_{\mathrm{ct}} \end{Bmatrix} = \begin{bmatrix} E_{\mathrm{n}} & 0 \\ 0 & E_{\mathrm{t}} \end{bmatrix} \left\{ \begin{Bmatrix} u_{\mathrm{cn}} \\ u_{\mathrm{ct}} \end{Bmatrix} - \begin{bmatrix} -1 & 0 & 0 \\ 0 & 1 & -1 \end{bmatrix} \begin{Bmatrix} \lambda_{\mathrm{c}1} \\ \lambda_{\mathrm{c}2} \\ \lambda_{\mathrm{c}3} \end{Bmatrix} \right\}. \tag{2.28a}$$

(b) In 3D:

$$\begin{Bmatrix} -p_{\mathrm{cn}} \\ p_{\mathrm{ct}_1} \\ p_{\mathrm{ct}_2} \end{Bmatrix} = \begin{bmatrix} E_{\mathrm{n}} & 0 & 0 \\ 0 & E_{\mathrm{t}} & 0 \\ 0 & 0 & E_{\mathrm{t}} \end{bmatrix} \left\{ \begin{Bmatrix} u_{\mathrm{cn}} \\ u_{\mathrm{ct}_1} \\ u_{\mathrm{ct}_2} \end{Bmatrix} - \sum_{k=2}^{9} \begin{bmatrix} -1 & 0 \\ 0 & \cos\beta_{k-1} \\ 0 & \sin\beta_{k-1} \end{bmatrix} \begin{Bmatrix} \lambda_{\mathrm{c}1} \\ \lambda_{\mathrm{c}k} \end{Bmatrix} \right\}. \tag{2.28b}$$

This explicit penalty formulation matrix form may be presented in a more compact unified relation valid for 2D and 3D problems:

$$\{p_{\mathrm{c}}\} = [D_{\mathrm{c}}]\{\{u_{\mathrm{c}}\} - [C_{\mathrm{c}}]\{\lambda_{\mathrm{c}}\}\}, \tag{2.29a}$$

where $\{p_{\mathrm{c}}\}$ are the normal and tangential contact tractions, $[D_{\mathrm{c}}]$ the elastic (stiffness) matrix, $\{u_{\mathrm{c}}\}$ the normal and tangential relative displacements, $[C_{\mathrm{c}}]$ the contact separation-slip (yield) matrix, and $\{\lambda_c\}$ the normal and tangential contact multipliers; and, in which:

(a) In 2D:

$$\{p_c\} = \begin{Bmatrix} -p_{cn} \\ p_{ct} \end{Bmatrix}, \quad [D_c] = \begin{bmatrix} E_n & 0 \\ 0 & E_t \end{bmatrix}, \quad \{u_c\} = \begin{Bmatrix} u_{cn} \\ u_{ct} \end{Bmatrix}, \quad [C_c] = \begin{bmatrix} -1 & 0 & 0 \\ 0 & 1 & -1 \end{bmatrix}, \quad \{\lambda_c\} = \begin{Bmatrix} \lambda_{c1} \\ \{\lambda_{c(k+1)}\} \end{Bmatrix}. \quad k \text{ is given by eqn (2.15b).} \tag{2.29b}$$

(b) In 3D:

$$\{p_c\} = \begin{Bmatrix} -p_{cn} \\ p_{ct_1} \\ p_{ct_2} \end{Bmatrix}, \quad [D_c] = \begin{bmatrix} E_n & 0 & 0 \\ 0 & E_t & 0 \\ 0 & 0 & E_t \end{bmatrix}, \quad \{u_c\} = \begin{Bmatrix} u_{cn} \\ u_{ct_1} \\ u_{ct_2} \end{Bmatrix}, \quad [C_c]\{\lambda_c\} = \sum_{k=2}^{9} \begin{bmatrix} -1 & 0 \\ 0 & \cos\beta_{k-1} \\ 0 & \sin\beta_{k-1} \end{bmatrix} \begin{Bmatrix} \lambda_{c1} \\ \lambda_{ck} \end{Bmatrix}. \quad k \text{ and } \beta_k \text{ are given by eqn(2.15b).} \tag{2.29c}$$

It is interesting to note that the normal and tangential penalty parameters, E_n and E_t, which actually define the elastic or stiffness coefficients of matrix $[D_c]$, represent the stiffness in the normal and tangential direction between each point of pairs considered in the potential contact surfaces. Here their connections can be interpreted as being modeled by a number of springs (fig. 2.9). On the other hand, E_n can be seen as the stiffness of the potential contact surface to penetration and E_t to tangential displacements.

Finally, substituting eqns (2.29a–c) into eqns (2.17a–c), penalty complementarity contact conditions may be written in the following form:

$$\{f_c\} = [C_\ell][D_c]\{\{u_c\} - [C_c]\{\lambda_c\}\} = [C_\ell][D_c]\{u_c\} - [C_\ell][D_c][C_c]\{\lambda_c\}, \tag{2.30a}$$

$$\begin{matrix} \{f_c\} \le 0, \\ \{\lambda_c\} \ge 0, \end{matrix} \quad \{f_c\}^T\{\lambda_c\} = 0 \quad \text{(complementarity condition).} \tag{2.30b}$$

It should be noted that $\{u_c\}$ and $\{\lambda_c\}$ are the unknowns and their determination requires the introduction of a system of supplementary equations. These equations can be derived by using a numerical discretization method of the boundary element formulation of the elastic contact problem described in Section 2.5.

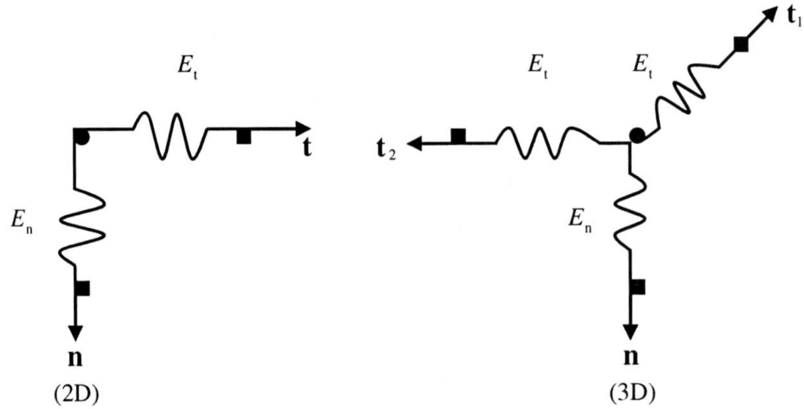

Figure 2.9: Physical representation of normal and tangential penalty parameters.

2.5 Boundary element formulation for elastic frictional contact problems

In the theory of linear elasticity, contact problems are one of the important practical topics that are inherently nonlinear and which occur on the potential contact boundaries. Due to the intrinsic boundary nature of both the contact problem and the boundary element method (BEM), this method has become one of the powerful formulations used to perform the analysis of frictional contact problems between deformable elastic bodies in engineering. This approach provides a set of equations relating contact boundary tractions and displacements with prescribed surface tractions and displacements for the contacting bodies. These relations then provide the additional equations required for the determination of the contact unknowns of eqns (2.30a,b).

In order to derive the boundary element formulation for the elastic contact problems, we start by presenting the boundary integral equations for each body. The presentation is based on the so-called direct approach and follows the notation used by Brebbia and Dominguez (1989). Let us then consider the contacting bodies before contact phenomena take place (fig. 2.1), and assume, for simplicity, that isotropic small elastic behavior is prevailing. Standard results from solid mechanics then lead to the following set of governing equations:

(1) Static equilibrium equations:

$$\sigma_{kj,j} = 0 \quad \text{in } V \text{ (without body force) and } k, j = 1, 2, 3, \tag{2.31}$$

where $\sigma_{kj,j}$ are the stress components.

(2) Boundary conditions:

(b1) Essential conditions or prescribed displacements:

$$u_k = \bar{u}_k \quad \text{on } S_u, \quad k = 1, 2, 3. \tag{2.32a}$$

(b2) Natural conditions or prescribed tractions:

$$t_k = \bar{t}_k \quad \text{on } S_f, \quad k = 1, 2, 3. \tag{2.32b}$$

(3) Displacement–strain relationships (compatibility or kinematic equations):

$$\varepsilon_{kj} = \frac{1}{2}(u_{k,j} + u_{j,k}) \quad \text{in } V \text{ and } k, j = 1, 2, 3, \tag{2.33}$$

where $u_{k,j}$ and ε_{kj} are the displacement derivatives and strain components, respectively.

(4) Strain–stress relationships (material constitutive law or generalized Hooke's law):

$$\sigma_{kj} = \frac{E}{(1+\nu)}\left[\frac{\nu}{(1-2\nu)}\delta_{kj}\varepsilon_{\ell\ell} + \varepsilon_{kj}\right] \quad \text{in } V \text{ and } k,\ j = 1, 2, 3, \tag{2.34a}$$

where E and ν are Young's modulus and Poisson's ratio, respectively, and δ_{kj} is the Kronecker delta symbol ($\equiv 1$ for $k = j$ and $\equiv 0$ for $k \neq j$). Furthermore, the strain $\varepsilon_{\ell\ell}$ can be referred to as the volumetric strain:

$$\varepsilon_{\ell\ell} = \varepsilon_{11} + \varepsilon_{22} + \varepsilon_{33}. \tag{2.34b}$$

Equations (2.31–2.34a,b) form a system of 15 equations (three equilibrium equations, six displacement–strain relations, and six strain–stress relations) to be used for the determination of the 15 unknowns (three displacements, six strains, and six stresses) involved in an elasticity problem.

It is instructive to write the strain–stress relationships in terms of displacements by substituting eqn (2.33) into eqns (2.34a,b):

$$\sigma_{kj} = \frac{E}{(1+\nu)}\left[\frac{\nu}{(1-2\nu)}\delta_{kj}u_{\ell,\ell} + \frac{1}{2}(u_{k,j} + u_{j,k})\right] \quad \text{in } V \text{ and } k,\ j = 1, 2, 3, \tag{2.35a}$$

where

$$u_{\ell,\ell} = u_{1,1} + u_{2,2} + u_{3,3}. \tag{2.35b}$$

Differentiating eqn (2.35a) and substituting into the stress equilibrium relation (eqn (2.31)) results in the well-known Navier's equation for elastostatics in terms of displacements:

$$u_{k,jj} + \frac{1}{(1-2\nu)}u_{j,kj} = 0 \quad \text{in } V \text{ and } k,\ j = 1, 2, 3. \tag{2.36}$$

All the equations given above are used for 3D analyses and for the plane strain case (k, $j = 1, 2$). When the plane stress condition is assumed (k, $j = 1, 2$), the same formulation may be maintained simply by replacing E by $E(1+2\nu)/(1+\nu)^2$ and ν by $\nu/(1+\nu)$.

Integral equation formulation for elasticity can be derived using the weighted residual method (Finlayson, 1972). By introducing a weighting function u_k^* (which are the virtual displacements), static equilibrium equations, eqn (2.31) can now be written in a weighted residual formulation using the following orthogonalized product (Brebbia, 1978):

$$\int_V (\sigma_{kj,j})u_k^* \, \mathrm{d}V = 0. \tag{2.37}$$

Integrating this expression, twice by parts, one finds:

$$\int_V \sigma_{kj,j}^* u_k \, \mathrm{d}V = \int_S t_k^* u_k \, \mathrm{d}S - \int_S u_k^* t_k \, \mathrm{d}S, \tag{2.38}$$

where t_k^* is the surface tractions corresponding to the u_k^* system. On the other hand, u_k^* and t_k^* are the so-called displacement and traction fundamental solutions and are chosen as the weighting functions. $\sigma_{kj,j}^*$ are also the stress fundamental solutions for the static equilibrium equations that we now look for.

The fundamental solution for the elasticity problem satisfies the following equation:

$$\sigma^*_{\ell j,j} + \Delta^i e_\ell = 0, \tag{2.39}$$

where $\Delta^i e_\ell$ is the Dirac delta function, represents a unit load at the point i in the ℓ direction. Therefore, the first term of eqn (2.38) becomes:

$$\int_V \sigma^*_{kj,j} u_k \, \mathrm{d}V = \int_V \sigma^*_{\ell j,j} u_\ell \, \mathrm{d}V = -\int_V \Delta^i e_\ell u_\ell \, \mathrm{d}V = -u^i_\ell e_\ell, \tag{2.40}$$

where u^i_ℓ represents the ℓth component of the displacements at the point i of application of the unit load.

Furthermore, the fundamental solutions u^*_k and t^*_k can be written in the form:

$$\begin{aligned} u^*_k &= u^*_{\ell k} e_\ell, \\ t^*_k &= t^*_{\ell k} e_\ell, \end{aligned} \tag{2.41}$$

where $u^*_{\ell k}$ and $t^*_{\ell k}$ are the k components of displacements and tractions, respectively, due to a unit concentrated load at the point i acting in the direction ℓ.

Substituting eqns (2.40, 2.41) into eqn (2.38) yields the integral equation for elastostatics, which is sometimes called the Somigliana's identity (without body force):

$$u^i_\ell + \int_S t^*_{\ell k} u_k \, \mathrm{d}S = \int_S u^*_{\ell k} t_k \, \mathrm{d}S. \tag{2.42}$$

This equation is valid for any point inside the body for any particular point where the unit forces are applied. When the boundary values u_k and t_k at all boundary points as well as the fundamental solutions $u^*_{\ell k}$ and $t^*_{\ell k}$ are known, the displacement components u^i_ℓ at any internal point i may be calculated by this integral equation.

The fundamental solutions $u^*_{\ell k}$ for a 3D isotropic body, derived from Navier's equations (2.36), are the so-called Kelvin's solution (Love, 1944):

$$u^*_{\ell k} = \frac{1}{16\pi G(1-\nu)r}\left[(3-4\nu)\delta_{\ell k} + \frac{r_\ell r_k}{r^2}\right], \tag{2.43}$$

where $G = E/2(1+\nu)$ is the shear modulus and the variable r represents the distance between the point i of application of the unit force and any other field point (observation point).

The traction fundamental solutions $t^*_{\ell k}$ corresponding to the Kelvin solution are:

$$t^*_{\ell k} = -\frac{1}{8\pi(1-\nu)r^2}\left[\frac{\partial r}{\partial n}\left[(1-2\nu)\delta_{\ell k} + \frac{3r_\ell r_k}{r^2}\right] + (1-2\nu)\left(n_\ell \frac{r_k}{r} - n_k \frac{r_\ell}{r}\right)\right], \tag{2.44}$$

where n_ℓ and n_k are the direction cosines of the normal vector $\mathbf{n}$ to the surface of the body with respect to ℓ and k directions. The normal derivative $\partial r/\partial \mathbf{n}$ is the derivative of the distance vector r with respect to the normal vector $\mathbf{n}$, and is equal to the inner product of the gradient of vector r and the normal vector $\mathbf{n}$.

Similarly, for the 2D isotropic plane strain case:

$$u^*_{\ell k} = \frac{1}{8\pi G(1-\nu)}\left[(3-4\nu)\ln\left(\frac{1}{r}\right)\delta_{\ell k} + \frac{r_\ell r_k}{r^2}\right], \tag{2.45}$$

$$t^*_{\ell k} = -\frac{1}{4\pi(1-\nu)r}\left[\frac{\partial r}{\partial \mathbf{n}}\left[(1-2\nu)\delta_{\ell k} + \frac{2r_\ell r_k}{r^2}\right] + (1-2\nu)\left(n_\ell \frac{r_k}{r} - n_k \frac{r_\ell}{r}\right)\right]. \tag{2.46}$$

Notice that in the 2D plane stress cases ν should be replaced by $\nu/(1+\nu)$.

We now attempt to transform the integral equation (2.42) into a boundary integral formulation by moving the internal point i to the boundary. In such a case, however, the integrals present different types of singularities depending on the considered kernel function, and one needs to analyze this behavior. This has been done in detail by Brebbia and Dominguez (1978, 1989), where it is shown that eqn (2.42) may be written in indicial notation:

$$c^i_{\ell k} u^i_k + \int_S t^*_{\ell k} u_k \, \mathrm{d}S = \int_S u^*_{\ell k} t_k \, \mathrm{d}S, \tag{2.47a}$$

where the symbol $c^i_{\ell k}$ is used to account for the local geometric characteristics at the location i of the unit concentrated load as follows:

$$c^i_{\ell k} = \begin{cases} 0; & i \text{ is outside } S, \\ \delta_{\ell k}; & i \text{ is inside } S \text{(internal point)}, \\ \frac{1}{2}\delta_{\ell k}; & i \text{ is on the smooth part of } S \text{(boundary point)}, \\ c^i_{\ell k}; & i \text{ is a corner point on } S. \end{cases} \tag{2.47b}$$

When point i is at a point where the boundary is non-smooth, Hartmannn (1980) has presented expressions for $c^i_{\ell k}$. However, explicit calculation is not usually necessary and can be obtained through the imposition of rigid-body displacements (Brebbia and Dominguez, 1989) or even computed implicitly by using a non singular formulation (Gakwaya, 1983).

It should be noted that, based on the above definitions for $c^i_{\ell k}$, the boundary integral formulation presented by eqns (2.47a,b) can be applied for points on the boundary as well as for internal points and thus allows one to solve the general elastic boundary value problems.

In order to solve eqns (2.47a,b) numerically, one can use the following matrix form:

$$[c]^i\{u\}^i + \int_S [t]^*\{u\} \, \mathrm{d}s = \int_S [u]^*\{t\} \, \mathrm{d}s, \tag{2.48a}$$

where

$$\{u\} = \left\{ \begin{Bmatrix} u_1 \\ u_2 \end{Bmatrix} \text{ in 2D}, \quad \begin{Bmatrix} u_1 \\ u_2 \\ u_3 \end{Bmatrix} \text{ in 3D} \right. \tag{2.48b}$$

is a vector whose components, u_k, are the displacements at any point on S.

$$\{u\}^i = \left\{ \begin{Bmatrix} u_1 \\ u_2 \end{Bmatrix}^i \text{ in 2D}, \quad \begin{Bmatrix} u_1 \\ u_2 \\ u_3 \end{Bmatrix}^i \text{ in 3D} \right. \tag{2.48c}$$

is a vector whose components, u^i, are the displacements at the internal or boundary point i where the load is applied.

$$\{t\} = \left\{ \begin{Bmatrix} t_1 \\ t_2 \end{Bmatrix} \text{in 2D,} \quad \begin{Bmatrix} t_1 \\ t_2 \\ t_3 \end{Bmatrix} \text{in 3D} \right. \tag{2.48d}$$

is a vector whose components, t_k, are the tractions at any point on S.

$$[t]^* = \left\{ \begin{bmatrix} t^*_{11} & t^*_{12} \\ t^*_{21} & t^*_{22} \end{bmatrix} \text{in 2D,} \quad \begin{bmatrix} t^*_{11} & t^*_{12} & t^*_{13} \\ t^*_{21} & t^*_{22} & t^*_{23} \\ t^*_{31} & t^*_{32} & t^*_{33} \end{bmatrix} \text{in 3D} \right. \tag{2.48e}$$

is a matrix whose coefficients, $t^*_{\ell k}$, are the forces in the k direction due to a unit force at the point i under consideration, acting in the ℓ direction.

$$[u]^* = \left\{ \begin{bmatrix} u^*_{11} & u^*_{12} \\ u^*_{21} & u^*_{22} \end{bmatrix} \text{in 2D,} \quad \begin{bmatrix} u^*_{11} & u^*_{12} & u^*_{13} \\ u^*_{21} & u^*_{22} & u^*_{23} \\ u^*_{31} & u^*_{32} & u^*_{33} \end{bmatrix} \text{in 3D} \right. \tag{2.48f}$$

is a matrix whose coefficients, $u^*_{\ell k}$, are the displacements in the k direction due to a unit force at the point i under consideration, acting in the ℓ direction.

$$[c]^i = \left\{ \begin{bmatrix} c_{11} & c_{12} \\ c_{21} & c_{22} \end{bmatrix}^i \text{in 2D,} \quad \begin{bmatrix} c_{11} & c_{12} & c_{13} \\ c_{21} & c_{22} & c_{23} \\ c_{31} & c_{32} & c_{33} \end{bmatrix}^i \text{in 3D} \right. \tag{2.48g}$$

is a matrix whose coefficients, $c^i_{\ell k}$, have a constant value depending on the local geometric shape at point i under consideration.

Notice that $[t]^*$ and $[u]^*$ are known and some $\{u\}^i$, $\{u\}$ and $\{t\}$ are the unknown variables to be determined. In order to find the unknown parameters, we need in general to integrate numerically the functions $[t]^*$ and $[u]^*$ by dividing the surface S into many small segments or boundary elements. These boundary elements for 2D can be constant, linear or quadratic, or higher order line segments, and for 3D they can be triangles or quadrilaterals of various orders. Integration is then performed over small sections of the boundary surface S, and their contributions added together to complete the discrete surface integrals in eqns (2.48a–g).

At this point, we are ready to derive the boundary element formulation for each of the isotropic elastic contacting bodies. To this effect it is customary to express the $\{u\}$ and $\{t\}$ in terms of the values of these variables at discrete points on the boundary, called nodes. We can write a set of interpolation functions, which define approximately the values of displacement and traction fields at any point on the element in terms of the nodal values at the nodes of the element. Over each element, $\{u\}$ and $\{t\}$ have the following representations:

$$\begin{aligned} \{u\} &= [\phi]\,\{u\}^j, \\ \{t\} &= [\phi]\,\{t\}^j, \end{aligned} \tag{2.49}$$

where $\{u\}^j$ and $\{t\}^j$ are the element nodal displacement and traction vectors of dimensions 2ME for 2D and 3ME for 3D, ME being the number of nodes on the element. The interpolation function matrix $[\phi]$ is an array of shape functions of dimensions 2(2ME) in 2D and 3(3ME) in 3D, i.e.,

$$[\phi] = \begin{bmatrix} \phi_1 & 0 & 0 & \phi_2 & 0 & 0 & \dots & \phi_M & 0 & 0 \\ 0 & \phi_1 & 0 & 0 & \phi_2 & 0 & \dots & 0 & \phi_M & 0 \\ 0 & 0 & \phi_1 & 0 & 0 & \phi_2 & \dots & 0 & 0 & \phi_M \end{bmatrix}. \tag{2.50}$$

The interpolation functions for any element can be found in any standard boundary element textbook, such as the one by Brebbia and Dominguez (1989). Note that these functions for $\{u\}$ and $\{t\}$ are generally different. However, here they are assumed to be identical.

Substituting now the above functions into eqns (2.48a–g) and discretizing the boundary S, one obtains the following discretized form for a nodal point (Brebbia and Dominguez, 1989):

$$[c]^i\{u\}^i + \sum_{j=1}^{\mathrm{NE}} \left[\int_{S_j} [t]^* \{\phi\} \,\mathrm{d}s \right] \{u\}^j = \sum_{j=1}^{\mathrm{NE}} \left[\int_{S_j} [u]^* \{\phi\} \,\mathrm{d}s \right] \{t\}^j , \tag{2.51}$$

where $\sum$ from $j = 1$ to NE indicates summations over all the NE elements on the surface S and S_j is the surface of the element j. Since vectors $\{u\}^j$ and $\{t\}^j$ can be considered to be constants of integration, they can be removed from the integrals, as shown in eqn (2.51).

Due to the difficulties in evaluating eqn (2.51) analytically, in particular if the elements are curved, a numerical integration scheme is used. Therefore, by employing the Gauss–Legendre quadrature method (Stroud and Secrest, 1966), the above equation can be transformed into a system of equations for node i as follows:

$$[c]^i\{u\}^i + \sum_{j=1}^{M} [\hat{H}]^{ij} \{u\}^j = \sum_{j=1}^{M} [G]^{ij} \{t\}^j , \tag{2.52}$$

where M is the number of nodes, $[\hat{H}]^{ij}$ and $[G]^{ij}$ are the interaction coefficients relating node i with all nodes on the surface S of the body, and $\{u\}^j$ and $\{t\}^j$ are the unknown displacements and tractions at nodes j.

Introducing the notation:

$$[H]^{ij} = \begin{cases} [\hat{H}]^{ij} & \text{if } i \neq j, \\ [\hat{H}]^{ij} + [c]^i & \text{if } i = j, \end{cases} \tag{2.53}$$

and substituting eqn (2.53) into eqn (2.52), we obtain:

$$\sum_{j=1}^{M} [H]^{ij} \{u\}^j = \sum_{j=1}^{M} [G]^{ij} \{t\}^j . \tag{2.54}$$

Finally, by applying the contributions $[H]^{ij}$ and $[G]^{ij}$ for all i nodes, eqn (2.54) can be written in a matrix form given by the following global system equations:

$$[H]\{u\} = [G]\{t\}, \tag{2.55}$$

where matrices $[H]$ and $[G]$ are called the influence matrices, and the terms inside them are the influence coefficients.

Notice that when the values of displacements and tractions at the surface are known, one can find the displacements and stresses at any internal point as well as stresses on the surfaces, which are required in many practical applications. These formulations are presented in Appendix A.

We now attempt to employ the boundary element formulation for the elastic contact problems, to obtain equations relating contact boundary tractions and displacements with prescribed surface tractions and displacements for contacting bodies. For each of the two contacting bodies

α ($\alpha = 1, 2$), as presented in Section 2.2, the boundary element formulation (eqn (2.55)) can be written:

$$[H]^{\alpha}\{u\}^{\alpha} = [G]^{\alpha}\{t\}^{\alpha}. \tag{2.56}$$

By applying the boundary conditions with respect to inside and outside the potential contact surface S_c and rearranging the corresponding columns and rows in the matrices $[H]$ and $[G]$, one finds (as in Gakwaya *et al.*, 1992):

$$[H \quad H_c]^{\alpha} \begin{Bmatrix} u \\ u_c \end{Bmatrix}_g^{\alpha} = [G \quad G_c]^{\alpha} \begin{Bmatrix} \bar{t} \\ p_c \end{Bmatrix}_g^{\alpha}, \tag{2.57}$$

where $\{u\}_g^{\alpha}$ and $\{\bar{t}\}_g^{\alpha}$ represent unknown and known data outside the potential contact surface S_c respectively, while $\{u_c\}_g^{\alpha}$ and $\{p_c\}_g^{\alpha}$ represent the unknown contact displacements and tractions respectively. The subscript 'g' indicates that the equation is expressed in the global coordinate system.

By inverting matrix $[H \quad H_c]^{\alpha}$ and multiplying the inverse with matrix $[G \quad G_c]^{\alpha}$, the global system equation (2.57) becomes:

$$\begin{Bmatrix} u \\ u_c \end{Bmatrix}_g^{\alpha} = \left[[H \quad H_c\ ]^{-1}\ [G \quad G_c\ ]\right]^{\alpha} \begin{Bmatrix} \bar{t} \\ p_c \end{Bmatrix}_g^{\alpha} = \begin{bmatrix} f_{oo} & f_{oc} \\ f_{co} & f_{cc} \end{bmatrix}^{\alpha} \begin{Bmatrix} \bar{t} \\ p_c \end{Bmatrix}_g^{\alpha}, \tag{2.58}$$

where $\begin{bmatrix} f_{oo} & f_{oc} \\ f_{co} & f_{cc} \end{bmatrix}^{\alpha}$ is the flexibility matrix and the subscripts 'o' and 'c' indicate outside and inside the potential contact zone, respectively.

Equation (2.58) immediately provides the required relation between the contact displacements and tractions and the external loading $\{\bar{t}\}_g^{\alpha}$, as well as an equation to find the displacement components outside the contact surface. One obtains for the contact displacements:

$$\{u_c\}_g^{\alpha} = [f_{co}]^{\alpha}\{\bar{t}\}_g^{\alpha} + [f_{cc}]^{\alpha}\{p_c\}_g^{\alpha} \tag{2.59}$$

and for the displacements outside the contact zone:

$$\{u\}_g^{\alpha} = [f_{oo}]^{\alpha}\{\bar{t}\}_g^{\alpha} + [f_{oc}]^{\alpha}\{p_c\}_g^{\alpha}. \tag{2.60}$$

Notice that once contact tractions $\{p_c\}_g^{\alpha}$ have been found, $\{u\}_g^{\alpha}$ may be obtained immediately by eqn (2.60).

The contact variables, as described in Section 2.2, were presented in a local coordinate system. It is thus necessary to transform them in eqn (2.59) into the local coordinate system by means of a transformation matrix [Tr] such that:

$$\begin{aligned} \{u_c\}_g^{\alpha} &= [\mathrm{Tr}]\,\{u_c\}^{\alpha}, \\ \{p_c\}_g^{\alpha} &= [\mathrm{Tr}]\,\{p_c\}^{\alpha}, \end{aligned} \tag{2.61a}$$

where

(a) In 2D:

$$[\mathrm{Tr}] = \begin{bmatrix} \cos\theta_c & \sin\theta_c \\ -\sin\theta_c & \cos\theta_c \end{bmatrix}. \tag{2.61b}$$

(b) In 3D:

$$[\mathrm{Tr}] = \begin{bmatrix} \cos\theta_c & 0 & \sin\theta_c \\ 0 & 1 & 0 \\ -\sin\theta_c & 0 & \cos\theta_c \end{bmatrix}, \tag{2.61c}$$

in which θ_c is the contact tangent angle between the x-axis and $\mathbf{t}$ in 2D and the x-axis and the tangent $\mathbf{t}_1$ in 3D (fig. 2.10).

Note that the local contact variables $\{u_c\}^\alpha$ and $\{p_c\}^\alpha$ are shown without subscript ℓ for simplicity. Therefore, for the contact point pairs, eqn (2.59) becomes:

$$\{u_c\}^\alpha = \{F\}^\alpha + [A]^\alpha \{p_c\}^\alpha, \tag{2.62a}$$

where

$$\begin{aligned} \{F\}^\alpha &= [\mathrm{Tr}]^{-1} \{f_{co}\}^\alpha \{\bar{t}\}_g^\alpha, \\ [A]^\alpha &= [\mathrm{Tr}]^{-1} \{f_{cc}\}^\alpha [\mathrm{Tr}]. \end{aligned} \tag{2.62b}$$

In accordance with the normal and tangential components for the contact variables, eqn (2.62a) may be expressed explicitly:

(a) In 2D:

$$\begin{Bmatrix} u_{cn} \\ u_{ct} \end{Bmatrix}^\alpha = \begin{Bmatrix} F_n \\ F_t \end{Bmatrix}^\alpha + [A_n \quad A_t]^\alpha \begin{Bmatrix} p_{cn} \\ p_{ct} \end{Bmatrix}^\alpha. \tag{2.62c}$$

(b) In 3D:

$$\begin{Bmatrix} u_{cn} \\ u_{ct_1} \\ u_{ct_2} \end{Bmatrix}^\alpha = \begin{Bmatrix} F_n \\ F_{t_1} \\ F_{t_2} \end{Bmatrix}^\alpha + \left[A_n \quad A_{t_1} \quad A_{t_2}\right]^\alpha \begin{Bmatrix} p_{cn} \\ p_{ct_1} \\ p_{ct_2} \end{Bmatrix}^\alpha. \tag{2.62d}$$

Substituting eqns (2.62a–d) into eqn (2.3), presented in Section 2.2, one finds the following required relation between the contact displacements and contact tractions, which are obtained by employing the boundary element formulation:

$$\{u_c\} = \{F_c\} + [A_c]\{p_c\}, \tag{2.63a}$$

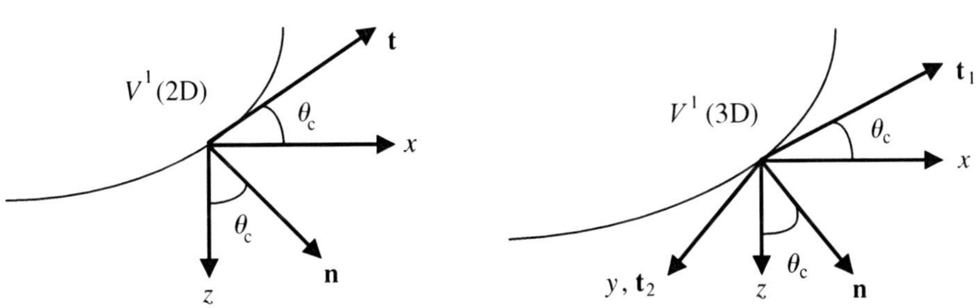

Figure 2.10: Local normal and tangential directions.

where

(a) In 2D:

$$
\begin{aligned}
\{u_c\} &= \begin{Bmatrix} u_{cn} \\ u_{ct} \end{Bmatrix} \equiv \begin{Bmatrix} u^2_{cn} - u^1_{cn} + u_0 \\ u^2_{ct} - u^1_{ct} \end{Bmatrix}, \\
\{F_c\} &= \begin{Bmatrix} F_{cn} \\ F_{ct} \end{Bmatrix} = \begin{Bmatrix} F^2_n - F^1_n + u_0 \\ F^2_t - F^1_t \end{Bmatrix}, \\
[A_c] &= [A_{cn} \quad A_{ct}] = \left[-(A^1_n + A^2_n) \quad (A^1_t + A^2_t)\right], \\
\{p_c\} &= \begin{Bmatrix} -p_{cn} \\ p_{ct} \end{Bmatrix}.
\end{aligned}
\tag{2.63b}
$$

(b) In 3D:

$$
\begin{aligned}
\{u_c\} &= \begin{Bmatrix} u_{cn} \\ u_{ct_1} \\ u_{ct_2} \end{Bmatrix} \equiv \begin{Bmatrix} u^2_{cn} - u^1_{cn} + u_0 \\ u^2_{ct_1} - u^1_{ct_1} \\ u^2_{ct_2} - u^1_{ct_2} \end{Bmatrix}, \\
\{F_c\} &= \begin{Bmatrix} F_{cn} \\ F_{ct_1} \\ F_{ct_2} \end{Bmatrix} = \begin{Bmatrix} F^2_n - F^1_n + u_0 \\ F^2_{t_1} - F^1_{t_1} \\ F^2_{t_2} - F^1_{t_2} \end{Bmatrix}, \\
[A_c] &= [A_{cn} \quad A_{ct}] = \left[-(A^1_n + A^2_n) \quad (A^1_{t_1} + A^2_{t_1}) \quad (A^1_{t_2} + A^2_{t_2})\right], \\
\{p_c\} &= \begin{Bmatrix} -p_{cn} \\ p_{ct_1} \\ p_{ct_2} \end{Bmatrix}.
\end{aligned}
\tag{2.63c}
$$

Thus, applying the boundary element formulation for elastic contact problems yields eqns (2.63a–c), which are the required supplementary equations needed to calculate the unknown contact variables presented in eqns (2.30a,b). The procedure to find them is explained in Section 2.6.

2.6 Linear complementarity formulation of elastic frictional contact problems

Here, attention is focused on the solution methodology of the elastic contact problems presented in the previous sections. According to Section 2.5, the important and necessary supplementary eqns (2.63a–c) are found to obtain the unknown contact variables.

Substituting eqns (2.29a–c) into these equations, a relation between the contact displacements $\{u_c\}$ and the contact multipliers $\{\lambda_c\}$ can be written:

$$\{u_c\} = \{E_c\} + [B_c]\,\{\lambda_c\}, \tag{2.64a}$$

where

$$
\begin{aligned}
\{E_c\} &= [[I] - [A_c]\,[D_c]]^{-1}\,\{F_c\}, \\
[B_c] &= -[[I] - [A_c]\,[D_c]]^{-1}\,[A_c]\,[D_c]\,[C_c],
\end{aligned}
\tag{2.64b}
$$

in which $[I]$ is the identity matrix.

Finally, substituting eqns (2.64a,b) into eqns (2.30a,b), penalty complementarity contact conditions yield the following linear complementarity problem (LCP) for the elastic contact problems:

$$-\{f_c\} = \{q_c\} + [M_c]\,\{\lambda_c\}\,, \tag{2.65a}$$

where

$$\begin{matrix} -\{f_c\} \geq 0, \\ \{\lambda_c\} \geq 0, \end{matrix} \qquad -\{f_c\}^{\mathrm{T}}\{\lambda_c\} = 0 \qquad \text{(complementarity condition)}, \tag{2.65b}$$

in which

$$\begin{aligned} \{q_c\} &= [C_\ell]\,[D_c]\,\{E_c\}\,, \\ [M_c] &= [C_\ell]\,[D_c]\,[[B_c] - [C_c]]\,. \end{aligned} \tag{2.65c}$$

The set of eqns (2.65a–c) is a standard linear complementarity problem whose general form is:

$$\{w\} = \{q\} + [M]\,\{z\}\,, \tag{2.66a}$$

$$\{w\} \geq 0, \quad \{z\} \geq 0 \qquad \text{(non-negativity conditions)}, \tag{2.66b}$$

$$\{w\}^{\mathrm{T}}\{z\} = 0 \qquad \text{(complementarity condition)}, \tag{2.66c}$$

where it is required to find vectors $\{w\}$, $\{z\}$ satisfying eqns (2.66a–c) for a given vector $\{q\}$ and a given matrix $[M]$.

This type of problem arises in a variety of fields, such as economics, mechanics and geometry and its properties have been studied widely as described by Murty (1988), Cottle *et al.* (1992) and Bazaraa *et al.* (1999). In recent years, the numerical solution of contact problems has been one of the important applications of the LCP techniques. If the contact boundary conditions can be expressed using non-negative multipliers, the problems can then be formulated in various forms of linear complementarity problems (LCP), a mathematical structure known from the area of mathematical programming, as shown in the present study by eqns (2.65a–c). There are several algorithms to solve the LCP (eqns (2.66a–c)), notably the principal pivoting method of Cottle and Dantzig (1963, 1968), the almost complementarity pivot algorithm or Lemke's algorithm of Lemke (1964, 1968), which is used in the present study, the conjugate gradient based projection method of Tamma *et al.* (1994) and the feasible interior point algorithm presented by Trinkle *et al.* (1997). Furthermore, there are two computer programs based on Lemke's algorithm. The first is used in our package and the second was examined by us to make sure that both gave the same results. The first program was implemented in FORTRAN IV by Ravindran (1972) and the second was provided by Michael Ferris and written in MATLAB (Trinkle *et al.*, 1997). A brief review of Lemke's algorithm is described in Appendix B.

By making use of Lemke's algorithm, $-\{f_c\}$ and $\{\lambda_c\}$ may be solved from eqns (2.66a–c) and consequently the unknown contact variables, $\{u_c\}$ and $\{p_c\}$, can be obtained from eqns (2.64a,b) and eqns (2.29a–c), respectively. Subsequently, the displacement components outside the contact surface (eqn (2.60)) as well as the displacements and stresses at any internal point and stresses on the surfaces can be determined, using the formulation presented in Appendix A.

Chapter 3

Three-dimensional elastic frictional contact analysis, Part 2: Applications

3.1 Introduction

In Chapter 2, following procedures based on concepts borrowed from the theory of elastoplasticity, the contact separation-slip (yield) functions corresponding to the unilateral normal contact and to the sliding threshold contact law have been derived, and using a penalty formulation, the contact problem has been reformulated in a penalty complementarity form. For numerical discretization purposes, the boundary element formulation for elastic contact problems was performed and yielded a discrete system of equations relating contact boundary tractions and displacements (contact variables) to the prescribed surface tractions and displacements. These relations were then coupled with the contact conditions through a penalty formulation. Finally, the coupled formulation was expressed as a linear complementarity problem (LCP) and solved by Lemke's algorithm.

The methodology described above has been implemented in a series of computer programs (about 100 subroutines), which form a package, called CONTACTA. This package is written in FORTRAN (66, 77, 90) and MATLAB (5, 6) languages and runs on a PC computer (Pentium/AMD). In this chapter, 2- and 3D elastic cases of frictional contact problems proposed in the package are considered. The boundary element codes of this package were first prepared by Gakwaya (1983) in his Ph.D. thesis, and were subsequently extended to the treatment of elastic contact problems through the works of Gakwaya *et al.* (1992) and Kong *et al.* (1992), which are based on the parametric LCP. On the other hand, the present study employs an LCP formulation which is easier to implement than the previous parametric LCP, and was successfully added to the package. While there is no limitation to the analysis of 2D elastic contact problems, certain limitations on the maximum number of nodes and elements must be applied in three-dimensions. Nevertheless, it is clear that by updating the package by fully exploiting the new FORTRAN (90, 95) language functionality, some limitations could be removed.

The macro-flow diagram of the package CONTACTA for the elastic case is shown in fig. 3.1. The main program defines the maximum dimensions of the system of equations and consists of three principal modules. The first module, i.e. Input data, has four block routines: BlCoor, BlBcnd, BlPrel and BlBlem. These routines read the coordinates of the boundary nodes, boundary conditions, material properties and element node connectivity respectively. The second module, BeTactMp, forms the system matrices $[H]$ and $[G]$ using a numerical integration scheme in the cases of singularity and non-singularity (Stcs and Stns) and rearranges them in accordance with

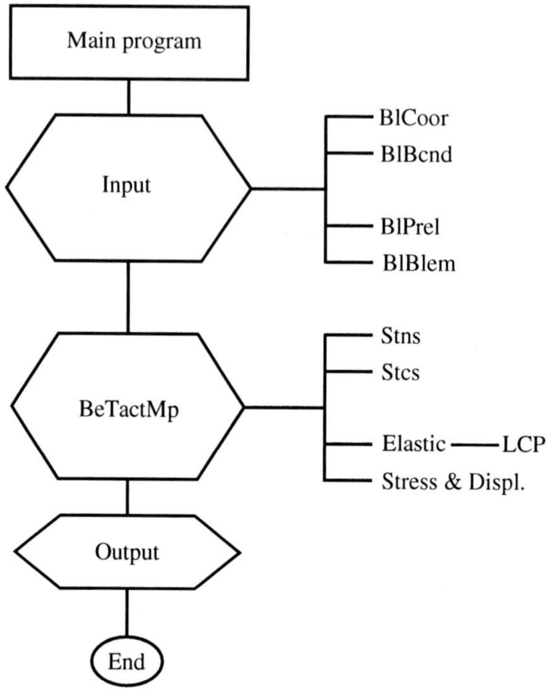

Figure 3.1: Macro-flow diagram of CONTACTA in elastic case.

the boundary conditions into a flexibility matrix in global as well as local coordinate systems, and subsequently performs the matrix partitioning required by contact problem penalty formulations (Elastic) in order to form the corresponding LCP. Finally, it calls on Lemke's algorithm (LCP) to calculate the contact variables, leading through to the last module (Output). Notice that by this routine the stresses and displacements for each node of contacting bodies can be obtained and outputted. In order to present the mesh generation of contacting bodies and illustrate the output results, a series of computer programs were written in MATLAB (5, 6) language.

In order to illustrate the scope of the package CONTACTA and give an indication of the effectiveness and accuracy of the proposed methodology and developments, the following series of numerical examples have been considered, and the results presented in four groups (11 examples):

1. Contact between a cylinder and a foundation.
2. Contact between a punch and a foundation (combined normal and tangential loading).
3. Contact between two cylinders.
4. Contact between a flat punch and a flat foundation (2D and 3D).

It should be noted that all the numerical results that appear in this chapter were obtained by systematically using double precision accuracy in the program, a single value of penalty parameters taken as $E_{\mathrm{n}} = E_{\mathrm{t1}} = E_{\mathrm{t2}} = 100E$, where E is Young's modulus, and quadratic isoparametric element approximated (3- and 8-noded). In order to compare our results, labeled CONTACTA in each figure, with those obtained from other analytical and numerical methods, we had to use the same body geometries, material properties and applied loadings. Because of the variety of units used in the published papers, these sometimes lead to rather awkward numerical values.

(Group 1)

3.2 Contact between a rigid cylinder and an elastic foundation

The first example is concerned with a Hertzian problem with friction that has been studied by Kikuchi and Oden (1984) using the finite element method (FEM). Under the assumption of plane strain conditions, a circular rigid cylinder is indented into the deformable foundation with the prescribed normal traction shown in fig. 3.2, which also illustrates the necessary characteristic values. For numerical purposes, the Young's modulus of the cylinder is taken to be 100 times that

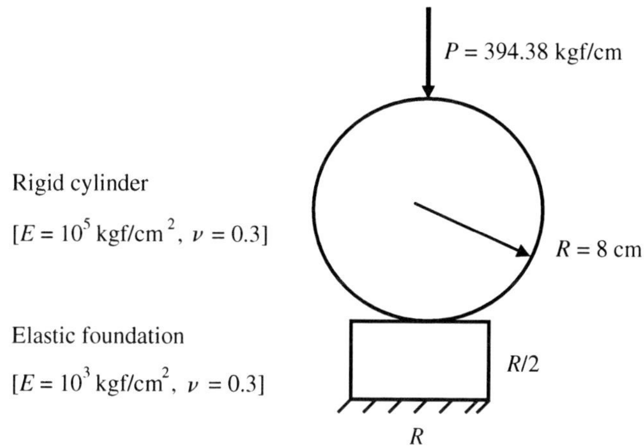

Figure 3.2: Contact between a circular rigid cylinder and an elastic foundation.

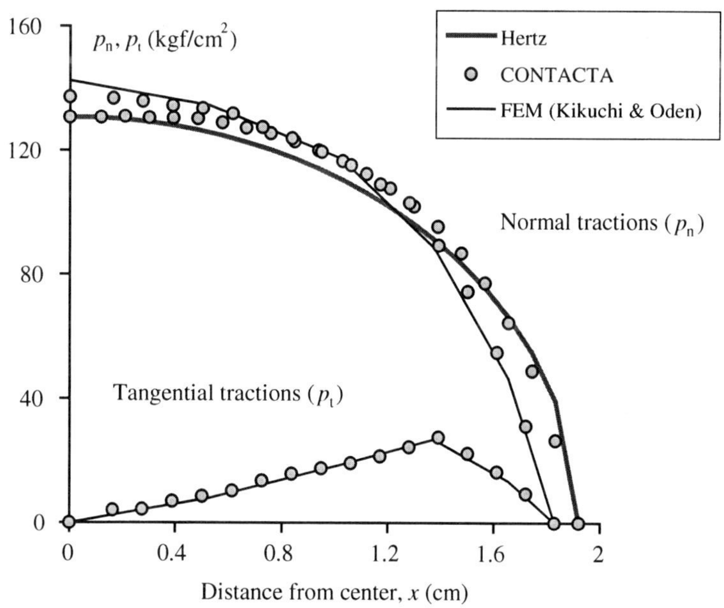

Figure 3.3: Distributions of the normal and tangential contact tractions for $\mu = 0.0$ and 0.3.

of the foundation in order to satisfy its rigidity assumption. Both the cylinder and the foundation have 48 quadratic elements, and 49 node pairs (24 quadratic elements) are considered in the potential contact zone. Computed results for two different coefficients of friction $\mu = 0.0$ and 0.3 are shown in fig. 3.3 and compared with the Hertz solution and Kikuchi and Oden's results (1984). Our results agree well with the analytical Hertz solution as well as the finite element solution of Kikuchi and Oden. From fig. 3.3, we see that the effect of friction is to reduce the contact length and to increase the maximum normal traction, a result that was also found by the FEM (Kikuchi and Oden, 1984).

3.3 Contact between an elastic cylinder and a rigid foundation

An elastic circular cylinder with radius $R = 60$ mm is pressed on a rigid rectangular cross sectional foundation with dimensions $(R \times R/2)$ under a normal load of 13,500 N/mm as shown in fig. 3.4.

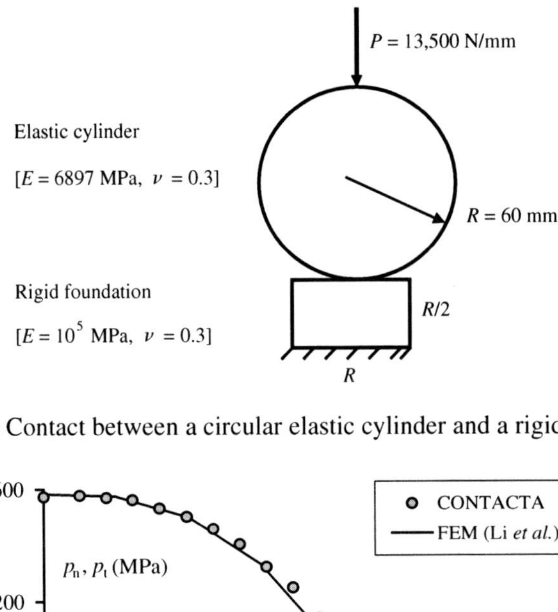

Figure 3.4: Contact between a circular elastic cylinder and a rigid foundation.

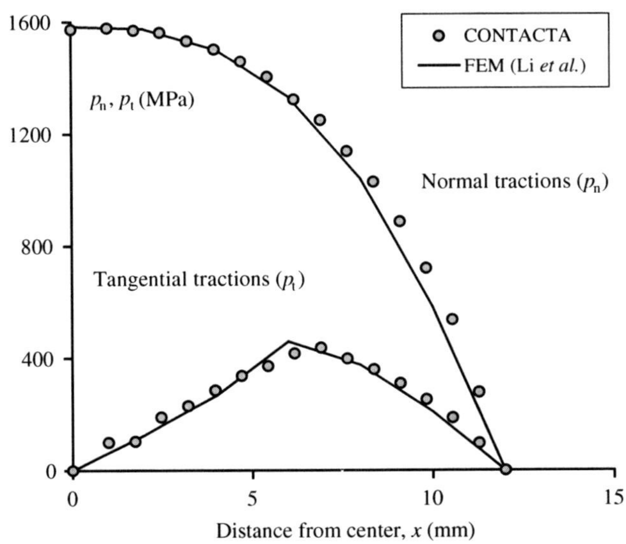

Figure 3.5: Distributions of the normal and tangential contact tractions for $\mu = 0.35$.

Both bodies have the same Poisson's ratio of 0.3, and Young's moduli are 6897 MPa and 10^5 MPa for the cylinder and the foundation respectively. This 2D contact problem has been studied using the finite element method (FEM) by Li *et al.* (1997) with the material properties being the reverse of those of the above example but using the same mesh arrangement in both studies. The normal and tangential tractions versus the length of the contact zone for a coefficient of friction $\mu = 0.35$ are illustrated in fig. 3.5 and are in good agreement with results obtained by Li *et al.* (1997) using the FEM.

3.4 Contact between a cylinder (circular disk) and a rectangular plate (foundation)

In this example, a rectangular plate (foundation) with dimensions ($4R \times 2R$) is compressed by a cylinder (circular disk) with radius $R = 50$ mm with a line force of 280.57 N/mm applied on the

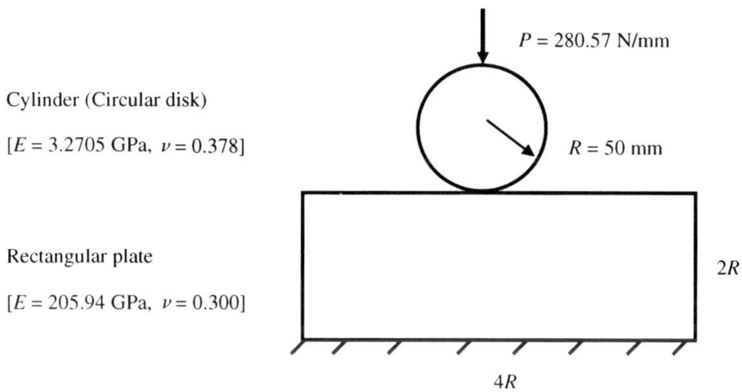

Figure 3.6: Contact between a cylinder (circular disk) and a rectangular plate.

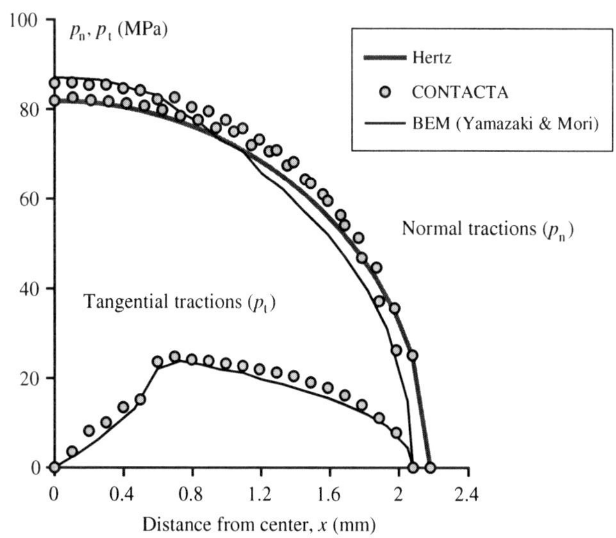

Figure 3.7: Distributions of the normal and tangential contact tractions for $\mu = 0.0$ and 0.3.

top of the cylinder as shown in fig. 3.6. Young's moduli and Poisson's ratios are 3.2705 GPa and 0.378 for the disk and 205.94 GPa and 0.300 for the plate. Assuming the same mesh generation scheme as in both previous examples for two contacting bodies and considering the plane strain situation, two different coefficients of friction, $\mu = 0.0$ and 0.3, were used and the normal pressure distribution as well as the tangential traction at the contact region were obtained and are illustrated in fig. 3.7. The results are compared for the case $\mu = 0.0$, with the classical Hertz solution and for the case $\mu = 0.3$, with the results obtained by Yamazaki and Mori (1988) who used the BEM and are seen to be in agreement with both results. The increase in maximum contact pressure and reduction in the contact zone due to friction are clearly observed in fig. 3.7, compared with the case of $\mu = 0.0$.

(Group 2)

3.5 Contact between an elastic punch and an elastic foundation (combined normal and tangential loading)

In order to study the influence of a prescribed tangential load on the contact status, we considered the contact between an elastic punch whose Young's modulus is 21 GPa and Poisson's ratio is

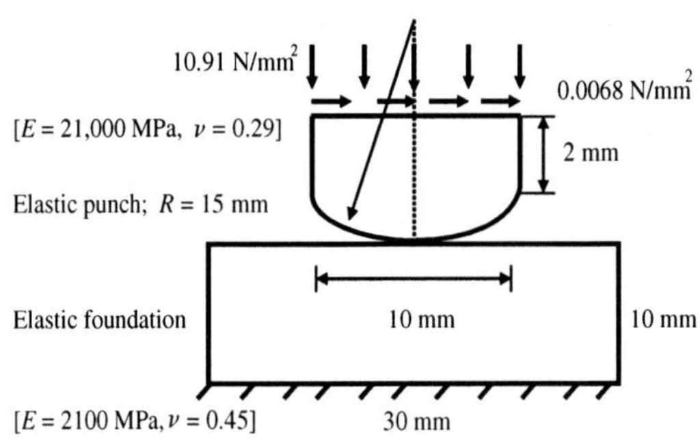

Figure 3.8: Contact between an elastic punch and an elastic foundation.

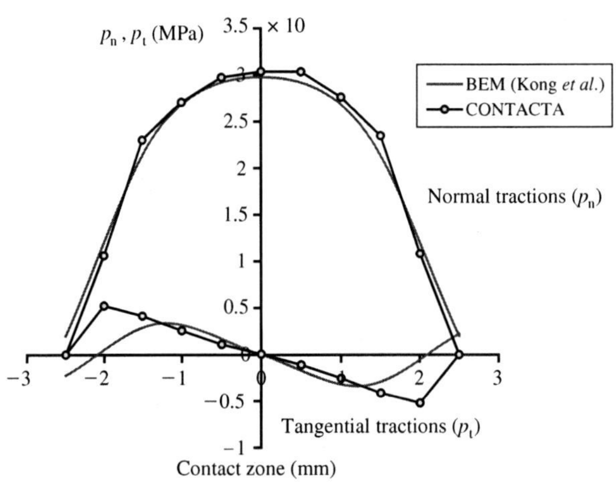

Figure 3.9: Distributions of the normal and tangential contact tractions for $\mu = 0.5$.

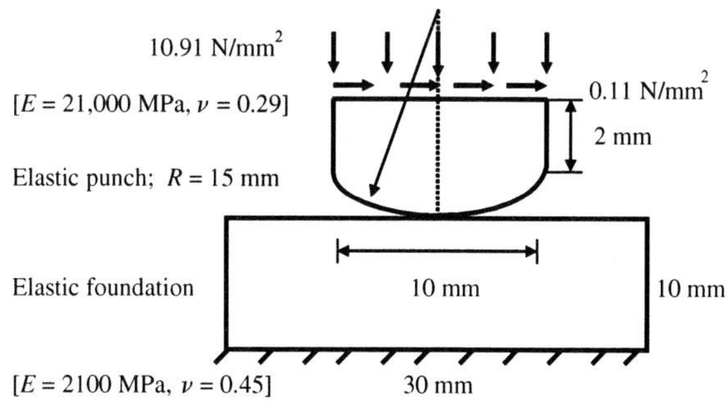

Figure 3.10: Contact between an elastic punch and an elastic foundation.

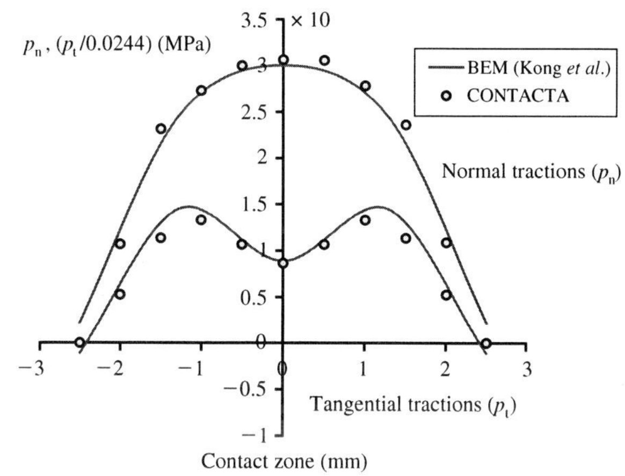

Figure 3.11: Distributions of the normal and tangential contact tractions for $\mu = 0.5$.

0.29 and an elastic foundation, which is characterized by a Young's modulus of 2.1 GPa and a Poisson's ratio of 0.45. A similar example can be found in the book of Kikuchi and Oden (1988) that applied FEM, and in the paper by Kong *et al.* (1992) that employed BEM. This is the case of plane strain with a coefficient of friction of 0.5 and the characteristic values of the model are illustrated in figs 3.8 and 3.10. The punch is indented into the foundation by a normal load of $10.91\,\mathrm{N/mm^2}$ while it is simultaneously subjected to a tangential load with two different values of 0.68×10^{-2} and $0.11\,\mathrm{N/mm^2}$, as shown in figs 3.8 and 3.10. The boundary of the punch is discretized into 16 quadratic elements, the foundation into 26 elements and the potential contact zone is defined by 13 node pairs (six quadratic elements). Computed results of the normal and tangential tractions distributed in the contact area for the two cases of tangential loading are given in figs 3.9 and 3.11 and compared with results obtained by Kong *et al.* (1992). The results show that the normal tractions seem to be insensitive to the values of the applied tangential load, whilst obviously the tangential tractions are significantly affected by changes in the magnitude of the tangential load. This phenomenon was also reported by Kikuchi and Oden (1988) and Kong *et al.* (1992). In addition, these distributions are slightly asymmetric with the centerline, which is consistent with the need to prevent rotation of the bodies under the action of the tangential loads. In order to obtain results similar to those of Kikuchi and Oden (1988), the same small values of prescribed tangential load were chosen and, as shown in figs 3.9 and 3.11, the results are in excellent agreement with those obtained in the aforementioned references. Notice that the results of this study are more accurate than those obtained by Kong *et al.* (1992), as illustrated in figs 3.9 and 3.11.

(Group 3)

3.6 Contact between two elastic cylinders

Two geometrically identical elastic cylinders, having the same material properties, are assumed to come into contact under an applied vertical compressive load as shown in figs 3.12 and

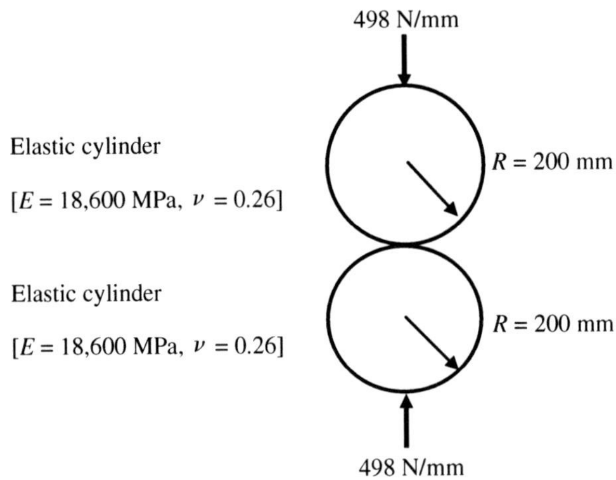

Figure 3.12: Contact between two elastic cylinders without friction.

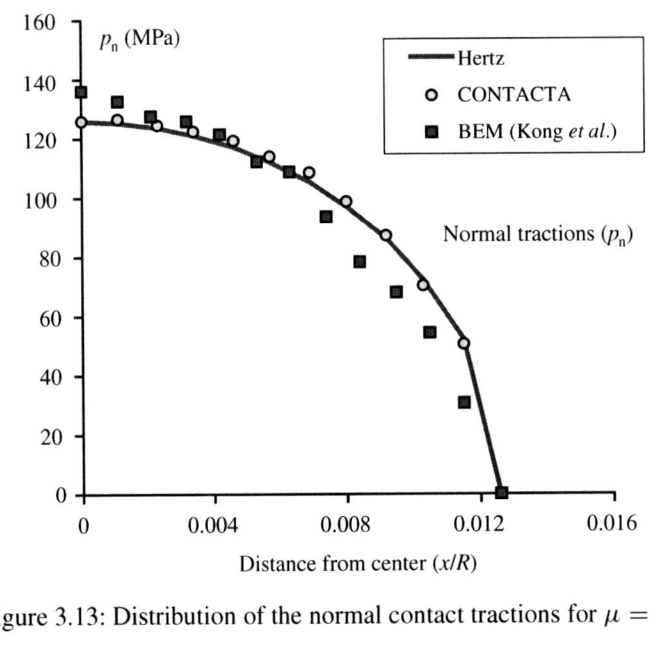

Figure 3.13: Distribution of the normal contact tractions for $\mu = 0.0$.

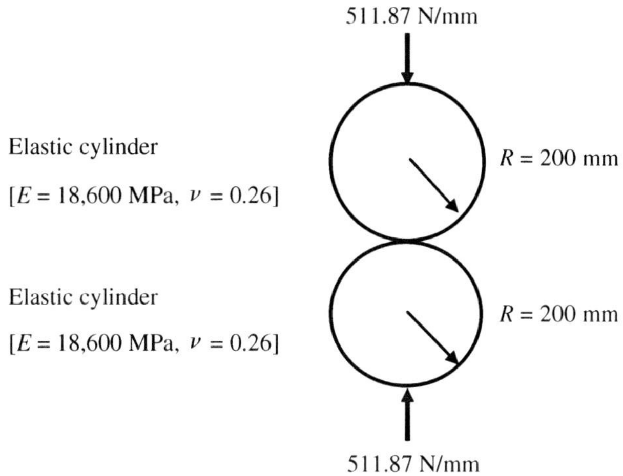

Figure 3.14: Contact between two elastic cylinders with friction.

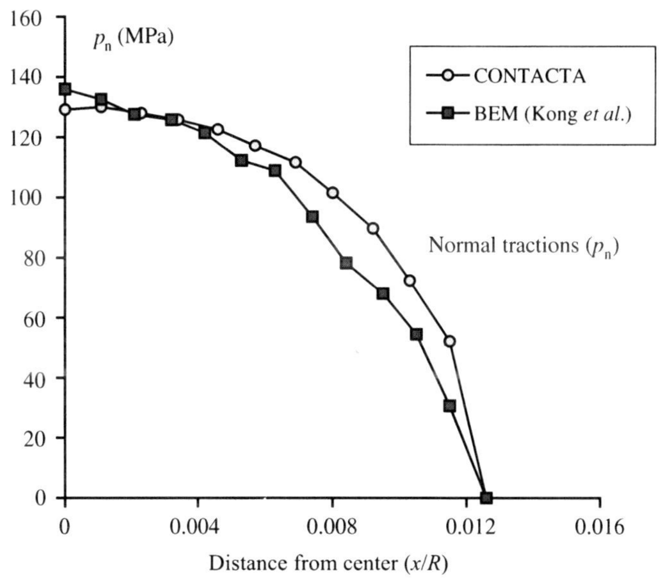

Figure 3.15: Distributions of the normal and tangential contact tractions for $\mu = 0.2$.

3.14. Both cylinders have been discretized into 48 quadratic elements, and 29 node pairs (14 quadratic elements) are considered in the potential contact zone. Under the assumption of plane strain and frictionless contact condition, i.e. $\mu = 0.0$, the distributed contact pressure obtained is shown in fig. 3.13 and compared with the classical Hertz solution and the results of Kong *et al.* (1992) based on the BEM (by correcting their scales as: $(x/R) \times 13.6$ and p_n (MPa)/13.6). In this case, the applied vertical load is taken as 498 N/mm and as illustrated in fig. 3.13, the results agree well with the Hertz solution. For the frictional contact condition, the coefficient of friction was taken as $\mu = 0.2$ and we used the same contact zone as in the frictionless case, with the value of the applied vertical load increased to 511.87 N/mm. Then the normal traction over the contact area was obtained and compared with the results of Kong *et al.* (1992) (by correcting their scales as: $(x/R) \times 13.6$ and p_n (MPa)/13.6), as shown in fig. 3.15. The agreement between the results in both the frictionless and frictional case is a good indication of the validity of the method and programming strategy. Notice that the effect of friction is to increase the value of the applied vertical load while the contact length remains the same as in the non-frictional case. It should be reported that very small and virtually negligible tangential tractions developed in the contact zone. This result agreed well with an analytical solution for two identical elastic cylinders for which the tangential tractions vanished (Johnson, 1985; Hills *et al.*, 1993).

(Group 4)

3.7 Contact between an elastic flat punch and an elastic flat foundation in 2D

The case of an elastic flat punch on an elastic flat foundation is found in many mechanical assemblies. This is a conforming contact problem, in which the contact area is known in advance.

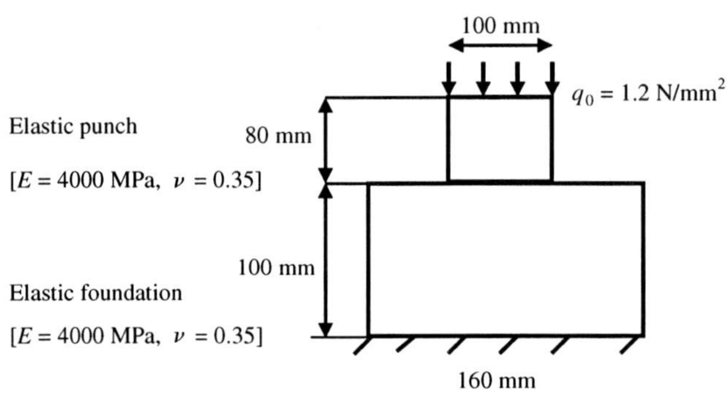

Figure 3.16: Contact between an elastic punch and an elastic foundation without and with friction.

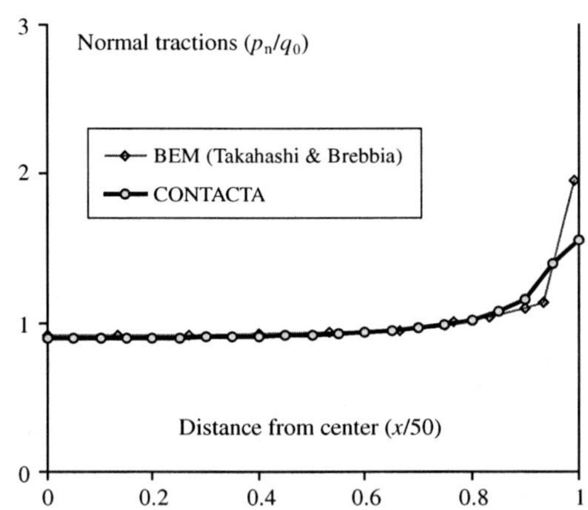

Figure 3.17: Distribution of the normal contact tractions for $\mu = 0.0$.

The geometry of the problem is shown in fig. 3.16. Both bodies are assumed to have a Poisson's ratio of 0.35, and a Young's modulus of 4.0 GPa. The punch and the foundation boundaries are divided into 42 and 56 quadratic elements, respectively, while 41 node pairs (20 quadratic elements) are considered as the potential contact zone. A uniform distributed traction, 1.2 N/mm^2, is applied to the top of the punch, and the foundation is assumed to rest on a smooth rigid surface. Two friction coefficients were tested, $\mu = 0.0$ and 0.2, and the problem solved under plane strain conditions. The normal contact pressure distributions for the two cases of friction, $\mu = 0.0$ and 0.2, were obtained and are shown in figs 3.17 and 3.18, respectively. The results are compared with boundary element solutions obtained by Takahashi and Brebbia (1988) employing the flexibility approach, by Paris and Garrido (1985) using discontinuous linear elements, and by Andersson (1982) applying continuous parabolic elements. The results given by the present study are in very

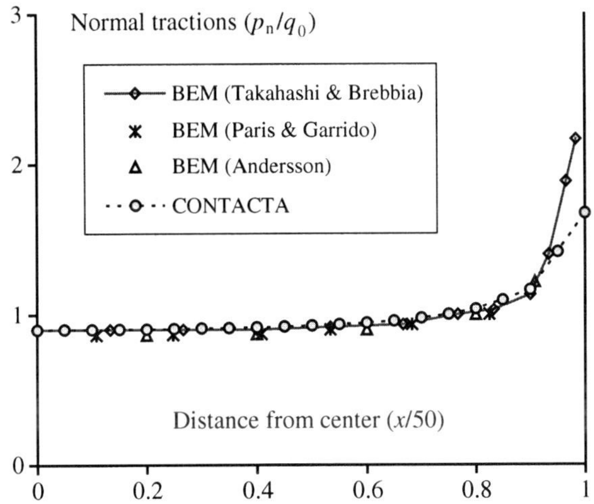

Figure 3.18: Distribution of the normal contact tractions for $\mu = 0.2$.

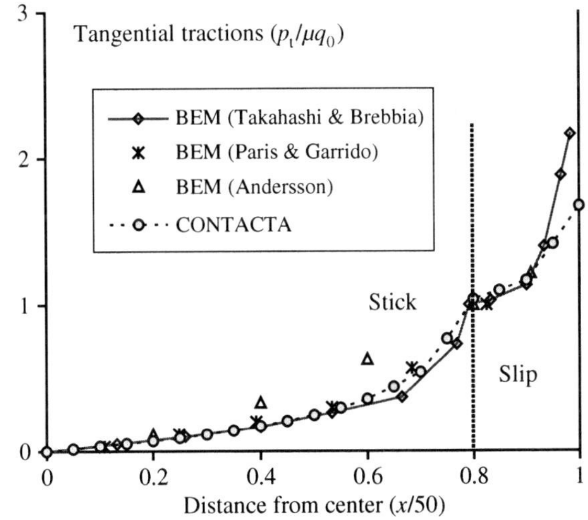

Figure 3.19: Distribution of the tangential contact tractions for $\mu = 0.2$.

good agreement with the other results and provide the values of normal traction near the end of the contact line as shown in the figures. Finally, the tangential tractions for the case of $\mu = 0.2$ were determined and compared with those results available in the literature as shown in fig. 3.19. The computed results are closer to the solutions of Takahashi and Brebbia (1988) and Paris and Garrido (1985), than to those of Andersson (1982). Furthermore, the present study yields the values of tangential traction near the end of the contact line.

3.8 Contact between an elastic flat punch and an elastic flat foundation in 3D

In order to demonstrate the ability and efficiency of the present study in dealing with 3D contact problems, the previous example, which was the contact between two elastic flat surfaces, has been treated in the 3D configuration shown in fig. 3.20. The boundary element mesh consists

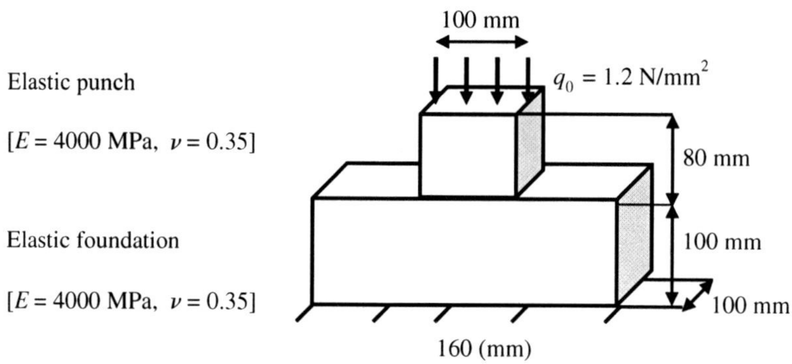

Figure 3.20: Contact between an elastic punch and an elastic foundation without and with friction in 3D.

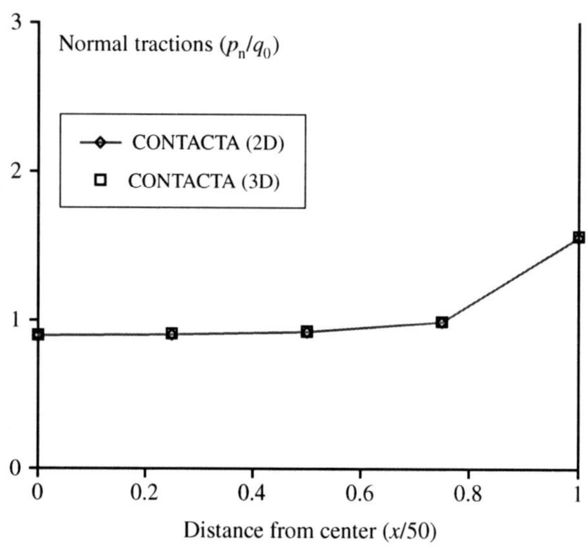

Figure 3.21: Distribution of the normal contact tractions for $\mu = 0.0$.

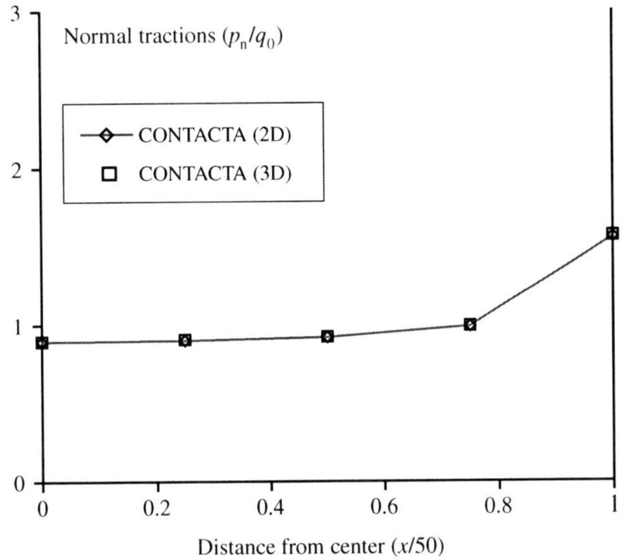

Figure 3.22: Distribution of the normal contact tractions for $\mu = 0.2$.

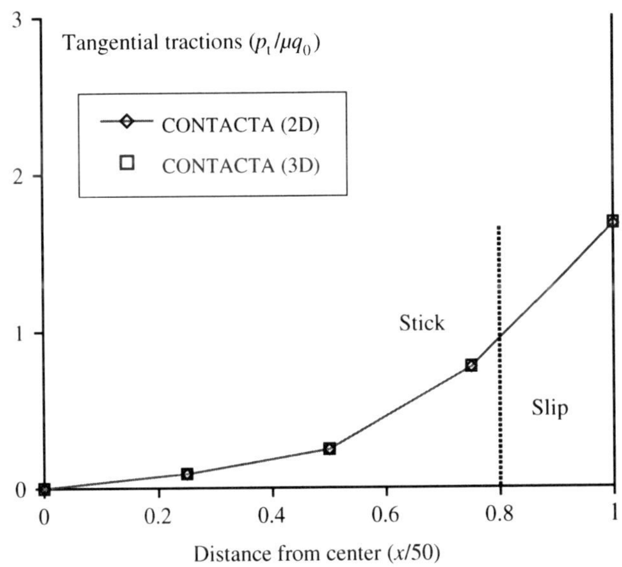

Figure 3.23: Distribution of the normal contact tractions for $\mu = 0.2$.

of 28 and 52 eight-node quadratic elements for the punch and foundation respectively, of which four are located within the contact area. Figure 3.21 shows the normal contact pressure for the frictionless case compared with the 2D calculations. Both numerical results in 2D and 3D are in very good agreement. For the frictional case, with $\mu = 0.2$, the normal and tangential contact tractions are obtained and compared with those of 2D calculations as illustrated in figs 3.22 and 3.23. Numerical results in 3D agree well with those in 2D.

CHAPTER 4

Three-dimensional elastoplastic frictional contact analysis, Part 1: Theory

4.1 Introduction

A general elastoplastic analysis method is presented for 2- and 3D contact problems between two deformable bodies undergoing small displacements with or without friction (Faraji *et al.*, 2003). A linear elastic perfectly plastic material behavior is assumed. The contact separation-slip and plastic yield functions are introduced and both the contact criterion and elastoplastic constitutive equation are expressed as a kind of linear complementarity problem (LCP). On the basis of these LCPs, and a boundary element formulation that utilizes some particular integrals (meshless method) to avoid volume integration, an incremental boundary element-complementarity method in the form of the unified linear complementarity problem is obtained to solve elastoplastic frictional contact problems. Finally, the mathematical programming problem is solved by Lemke's algorithm and the analysis successfully implemented through the package CONTACTA. Using this package, with the material assumed to obey the von Mises or Tresca criterion with their associated flow rule, one can obtain the value of the load at the first yield for which the first node becomes plastic, as well as stress and displacement components for each node of the contacting bodies. Furthermore, the contact zone extension, distributions of normal and tangential contact tractions and the resulting zone of micro-slip can be determined. Indeed, not only is this chapter a direct extension of Chapter 2 treating the elastic contact problems, but it can also be viewed as an LCP formulation for the elastoplastic frictional contact problem. This approach uses an elastic perfectly plastic type of friction law rather than the more complex parametric LCP presented by Kong *et al.* (1993), which was based on a rigid plastic model of friction law. It should be noted that a finite element method (FEM) model of this analysis in two dimensions has already been presented by Zhu (1995a).

4.2 Incremental description of penalty complementarity contact boundary conditions

In Chapter 2, contact conditions have been derived in a penalty complementarity form following concepts from the theory of elastoplasticity. The contact separation-slip (yield) functions were found for the normal unilateral and sliding threshold contact laws, adopting a penalty formulation using penalty contact parameters. This was first derived assuming an elastic material behavior and presented in a non-incremental form. However, for elastoplastic material behavior, in which the problem solution is dependent on the loading path, it becomes necessary, in general, to use the differential or incremental (rate) formulation presented in Chapter 2. Therefore, using the concepts and formulations in Chapter 2, the penalty complementarity contact conditions are presented in incremental form in the following analyses.

Considering the ith incremental step, the independent contact variables, i.e. contact tractions and relative contact displacements vectors, can be written:

$$\{p_c\}_i = \{p_c\}_{i-1} + \{\mathrm{d}p_c\}_i, \tag{4.1}$$

$$\{u_c\}_i = \{u_c\}_{i-1} + \{\mathrm{d}u_c\}_i, \tag{4.2}$$

where $\{p_c\}_{i-1}$ and $\{u_c\}_{i-1}$ are the known contact variables at the previous increment $(i-1)$.

Substituting eqns (4.1) and (4.2) into eqn (2.29a), and replacing $\{\lambda_c\}$ by $\{\mathrm{d}\lambda_c\}$, corresponding to the ith incremental step, we have:

$$\{\mathrm{d}p_c\}_i = [D_c]\{\{\mathrm{d}u_c\}_i - [C_c]\{\mathrm{d}\lambda_c\}_i\} + \{[D_c]\{u_c\}_{i-1} - \{p_c\}_{i-1}\}, \tag{4.3}$$

where matrices $[D_c]$ and $[C_c]$ are given by eqns (2.29b,c) for the 2D and 3D cases respectively.

In order to derive the incremental form of the penalty complementarity contact conditions, by substituting eqn (4.2) into eqns (2.30a,b), and replacing $\{\lambda_c\}$ by $\{\mathrm{d}\lambda_c\}$ with respect to the ith incremental step, one finally finds:

$$\begin{aligned}\{f_c\}_i &= [C_\ell][D_c]\{\{\mathrm{d}u_c\}_i - [C_c]\{\mathrm{d}\lambda_c\}_i\} + \{[C_\ell][D_c]\{u_c\}_{i-1}\} \\ &= [C_\ell][D_c]\{\mathrm{d}u_c\}_i - [C_\ell][D_c][C_c]\{\mathrm{d}\lambda_c\}_i + \{[C_\ell][D_c]\{u_c\}_{i-1}\},\end{aligned} \tag{4.4}$$

where

$$\begin{matrix}\{f_c\}_i \le 0, \\ \{\mathrm{d}\lambda_c\}_i \ge 0,\end{matrix} \quad \{f_c\}_i^{\mathrm{T}}\{\mathrm{d}\lambda_c\}_i = 0 \qquad \text{(complementarity condition)}, \tag{4.5}$$

where matrix $[C_\ell]$ is still given by eqns (2.17b,c) for 2D and 3D respectively. And, as described in Chapter 2, vectors $\{\mathrm{d}u_c\}_i$ and $\{\mathrm{d}\lambda_c\}_i$ are unknown in the current incremental step and one requires another system of supplementary equations to solve eqns (4.4, 4.5). These equations can be derived by employing a standard elastoplastic flow problem, as well as a numerical discretization method, such as the boundary element method described in the following sections.

4.3 Elastoplastic constitutive equations

Because the solution of plasticity problems depends on the loading history, a standard elastoplastic flow problem for each of the contacting bodies is governed by the following incremental relations (Kachanov, 1971; Martin, 1975; Maier *et al.*, 1982, 1983, 1987; Lemaitre and Chaboche, 1988; Zhong and Wu, 1993; Chen and Saleeb, 1994).

(1) Plastic yield function (yield condition):

$$f_p(\sigma_{ij}, \varepsilon_{ij}^{\mathrm{p}}, h) \le 0 \qquad i, j = 1, 2, 3. \tag{4.6}$$

(2) Plastic flow rule:

$$\mathrm{d}\varepsilon_{ij}^{\mathrm{p}} = \frac{\partial g_{\mathrm{p}}}{\partial \sigma_{ij}} \mathrm{d}\lambda_{\mathrm{p}} \qquad i, j = 1, 2, 3, \tag{4.7}$$

where the following apply.

(3) Complementarity conditions:

$$\mathrm{d}\lambda_{\mathrm{p}} \begin{cases} = 0, & \text{when } f_{\mathrm{p}} \leq 0, \\ \geq 0, & \text{when } f_{\mathrm{p}} = 0 \end{cases} \tag{4.8}$$

and σ_{ij}, $\mathrm{d}\varepsilon_{ij}^{\mathrm{p}}$ and h denote, respectively the stress tensor, the plastic strain increment vector, and the hardening parameter.

Inequality (4.6) expresses the plastic yield function (plastic yield condition), i.e. the stress resistance or strength, which defines in the stress space the current elastic domain whose boundary is the current yield surface or yield locus. If a stress point lies inside the surface, i.e. $f_{\mathrm{p}} < 0$, we call it an elastic state and only elastic behavior is expected at this stress point. On the other hand, for a state of stress lying on the yield surface, we have $f_{\mathrm{p}} = 0$ and it is referred to as a plastic state. Either the elastic or elastoplastic behavior may occur.

Equation (4.7) introduces the plastic flow rule, which defines a ratio among the components of $\mathrm{d}\varepsilon_{ij}^{\mathrm{p}}$, or the direction of the plastic strain increment vector. The plastic potential g_{p} is a function of the current stress states and plastic history, but is independent of incremental quantities and is defined for a given yield function. The gradient vector $\partial g_p/\partial\sigma_{ij}$ defines the direction of the plastic strain increment vector $\mathrm{d}\varepsilon_{ij}^{\mathrm{p}}$, i.e. normal to the potential surface $g_{\mathrm{p}} = 0$ at the current stress point. Consequently, this flow rule is called the normality condition. Finally, $\mathrm{d}\lambda_{\mathrm{p}}$ is called the plastic flow parameter or plastic multiplier and represents a measure of the amount of plastic flow. It is a non-negative scalar variable that will vary throughout the plastic loading history.

Equation (4.8) expresses the consistency condition, which postulates that yielding ($\mathrm{d}\lambda_{\mathrm{p}} > 0$) can only occur when the stress point stays on the current yield surface ($f_{\mathrm{p}} = 0$).

In order to derive a linear complementarity form of the above elastoplastic constitutive equations, the plastic yield function (yield condition) (eqn (4.6)) may be approximated by the linear terms of the Taylor's series at the current state:

$$f_{\mathrm{p}} = f_0 + \mathrm{d}f_{\mathrm{p}} = f_0 + \frac{\partial f_{\mathrm{p}}}{\partial \sigma_{ij}} \mathrm{d}\sigma_{ij} + \frac{\partial f_{\mathrm{p}}}{\partial \varepsilon_{ij}^{\mathrm{p}}} \mathrm{d}\varepsilon_{ij}^{\mathrm{p}} + \frac{\partial f_{\mathrm{p}}}{\partial h} \mathrm{d}h \leq 0, \tag{4.9}$$

where f_0 is the known value of f_{p} at the beginning of the incremental step and is the so-called yield limit in uniaxial loading ($f_0 = -Y$). The quantities $\partial f_{\mathrm{p}}/\partial\sigma_{ij}$, $\partial f_{\mathrm{p}}/\partial\varepsilon_{ij}^{\mathrm{p}}$, $\partial f_{\mathrm{p}}/\partial h$, in the above equation and $\partial g_{\mathrm{p}}/\partial\sigma_{ij}$ in eqn (4.7) can be considered as constant during the incremental step, and regarding the hardening parameter h, it is assumed (Hill, 1983; Chen and Saleeb, 1994) that the differential (increment) of h is a linear function of the plastic flow parameter $\mathrm{d}\lambda_{\mathrm{p}}$:

$$\mathrm{d}h = H\mathrm{d}\lambda_{\mathrm{p}}, \tag{4.10}$$

in which H is a constant related to the history of the plastic hardening.

Assuming a linear elastic perfectly plastic behavior of the material and the proposed associated flow rule, the plastic yield function (eqn (4.9)) simplifies to

$$f_{\mathrm{p}} = -Y + \frac{\partial f_{\mathrm{p}}}{\partial \sigma_{ij}} \mathrm{d}\sigma_{ij} \leq 0 \tag{4.11}$$

and the plastic flow rule (eqn (4.7)) in an associated formulation whereby the plastic potential surface has the same shape as the plastic yield surface, i.e. $g_\mathrm{p} = f_\mathrm{p}$, can be written:

$$d\varepsilon_{ij}^{\mathrm{p}} = \frac{\partial f_\mathrm{p}}{\sigma_{ij}} d\lambda_\mathrm{p} \tag{4.12}$$

and finally the complementarity conditions (eqn (4.8)) can be equivalently expressed:

$$\begin{matrix} f_\mathrm{p} \leq 0, \\ d\lambda_\mathrm{p} \geq 0, \end{matrix} \quad f_\mathrm{p} d\lambda_\mathrm{p} = 0. \tag{4.13}$$

Notice that eqns (4.11–4.13) represent the linear complementarity form of the constitutive laws of classically perfect or ideal plasticity.

For numerical computations, the plastic yield function f_p in eqns (4.11–4.13) can be approximated by introducing piecewise linear yield functions. On each approximating convex surface, these equations become:

$$f_{\mathrm{p}r} = -Y + \frac{\partial f_{\mathrm{p}r}}{\partial \sigma_{ij}} d\sigma_{ij}, \quad r = 1, \ldots, m_\mathrm{p}, \tag{4.14}$$

$$d\varepsilon_{ij}^{\mathrm{p}} = \sum_{r=1}^{m_\mathrm{p}} \frac{\partial f_{\mathrm{p}r}}{\partial \sigma_{ij}} d\lambda_{\mathrm{p}r}, \quad r = 1, \ldots, m_\mathrm{p}, \tag{4.15}$$

$$\begin{matrix} f_{\mathrm{p}r} \leq 0, \\ d\lambda_{\mathrm{p}r} \geq 0, \end{matrix} \quad f_{\mathrm{p}r} d\lambda_{\mathrm{p}r} = 0, \quad r = 1, \ldots, m_\mathrm{p}, \tag{4.16}$$

where $f_{\mathrm{p}r}$ are affine functions of σ_{ij}, and m_p is the number of approximating yield functions (planes) or yield modes in the principal stress state, respectively, equal to 6 and 8 for the Tresca and von Mises yield condition in the present study.

From the results of the previous section, in the stress space, σ_{ij}, at the ith incremental step, is given by:

$$\{\sigma\}_i = \{\sigma\}_{i-1} + \{d\sigma\}_i, \tag{4.17}$$

where $\{\sigma\}_i = \{\sigma_{11}\ \sigma_{22}\ \sigma_{12}\}_i^\mathrm{T}$ in 2D and $\{\sigma\}_i = \{\sigma_{11}\ \sigma_{22}\ \sigma_{33}\ \sigma_{12}\ \sigma_{23}\ \sigma_{31}\}_i^\mathrm{T}$ in 3D. Notice that in plane strain $\sigma_{33} = \nu(\sigma_{11} + \sigma_{22})$ and in plane stress $\sigma_{33} = 0$.

Substituting eqn (4.17) into eqns (4.14–4.16) leads to the linear complementarity problem of the constitutive laws of ideal plasticity in the following matrix form:

$$\{f_\mathrm{p}\}_i = \{\{-Y\} + [C_\mathrm{p}]_{i-1}\{\sigma\}_{i-1}\} + [C_\mathrm{p}]_{i-1}\{d\sigma\}_i, \tag{4.18}$$

$$\{d\varepsilon^\mathrm{p}\}_i = [C_\mathrm{p}]_{i-1}^\mathrm{T}\{d\lambda_\mathrm{p}\}_i, \tag{4.19}$$

$$\begin{matrix} \{f_\mathrm{p}\}_i \leq 0, \\ \{d\lambda_\mathrm{p}\}_i \geq 0, \end{matrix} \quad \{f_\mathrm{p}\}_i^\mathrm{T}\{d\lambda_\mathrm{p}\}_i = 0, \tag{4.20}$$

where

$$\begin{aligned} \{f_\mathrm{p}\}_i &= \{f_{\mathrm{p}1}\ \mathrm{f}_{\mathrm{p}2} \ldots \mathrm{f}_{\mathrm{p}m_\mathrm{p}}\}_i^\mathrm{T}, \\ \{-Y\} &= \{-Y - Y \ldots - Y\}^\mathrm{T}, \\ \{d\lambda_\mathrm{p}\}_i &= \{d\lambda_{\mathrm{p}1} d\lambda_{\mathrm{p}2} \ldots d\lambda_{\mathrm{p}m_\mathrm{p}}\}_i^\mathrm{T} \end{aligned} \tag{4.21}$$

and, in which $[C_\mathrm{p}]_{i-1}$ is called the plastic yield matrix, i.e. the matrix of outward unit normal vectors to the yield functions (planes), that is given below.

(a) In 2D:

$$\{\mathrm{d}\sigma\}_i = \{\mathrm{d}\sigma_{11} \quad \mathrm{d}\sigma_{22} \quad \mathrm{d}\sigma_{12}\}_i^{\mathrm{T}}, \tag{4.22a}$$

$$\{\mathrm{d}\varepsilon^{\mathrm{p}}\}_i = \{\mathrm{d}\varepsilon_{11}^{\mathrm{p}} \quad \mathrm{d}\varepsilon_{22}^{\mathrm{p}} \quad \mathrm{d}\varepsilon_{12}^{\mathrm{p}}\}_i^{\mathrm{T}}, \tag{4.22b}$$

$$[C_{\mathrm{p}}]_{i-1} = \begin{bmatrix} \dfrac{\partial f_{\mathrm{p}1}}{\partial \sigma_{11}} & \dfrac{\partial f_{\mathrm{p}1}}{\partial \sigma_{22}} & \dfrac{\partial f_{\mathrm{p}1}}{\partial \sigma_{12}} \\ \dfrac{\partial f_{\mathrm{p}2}}{\partial \sigma_{11}} & \dfrac{\partial f_{\mathrm{p}2}}{\partial \sigma_{22}} & \dfrac{\partial f_{\mathrm{p}2}}{\partial \sigma_{12}} \\ \cdots & \cdots & \cdots \\ \dfrac{\partial f_{\mathrm{p}m_{\mathrm{p}}}}{\partial \sigma_{11}} & \dfrac{\partial f_{\mathrm{p}m_{\mathrm{p}}}}{\partial \sigma_{22}} & \dfrac{\partial f_{\mathrm{p}m_{\mathrm{p}}}}{\partial \sigma_{12}} \end{bmatrix}_{i-1}. \tag{4.22c}$$

(b) In 3D:

$$\{\mathrm{d}\sigma\}_i = \{\mathrm{d}\sigma_{11} \quad \mathrm{d}\sigma_{22} \quad \mathrm{d}\sigma_{33} \quad \mathrm{d}\sigma_{12} \quad \mathrm{d}\sigma_{23} \quad \mathrm{d}\sigma_{31}\}_i^{\mathrm{T}}, \tag{4.23a}$$

$$\{\mathrm{d}\varepsilon^{\mathrm{p}}\}_i = \{\mathrm{d}\varepsilon_{11}^{\mathrm{p}} \quad \mathrm{d}\varepsilon_{22}^{\mathrm{p}} \quad \mathrm{d}\varepsilon_{33}^{\mathrm{p}} \quad \mathrm{d}\varepsilon_{12}^{\mathrm{p}} \quad \mathrm{d}\varepsilon_{23}^{\mathrm{p}} \quad \mathrm{d}\varepsilon_{31}^{\mathrm{p}}\}_i^{\mathrm{T}}, \tag{4.23b}$$

$$[C_{\mathrm{p}}]_{i-1} = \begin{bmatrix} \dfrac{\partial f_{\mathrm{p}1}}{\partial \sigma_{11}} & \dfrac{\partial f_{\mathrm{p}1}}{\partial \sigma_{22}} & \dfrac{\partial f_{\mathrm{p}1}}{\partial \sigma_{33}} & \dfrac{\partial f_{\mathrm{p}1}}{\partial \sigma_{12}} & \dfrac{\partial f_{\mathrm{p}1}}{\partial \sigma_{23}} & \dfrac{\partial f_{\mathrm{p}1}}{\partial \sigma_{31}} \\ \dfrac{\partial f_{\mathrm{p}2}}{\partial \sigma_{11}} & \dfrac{\partial f_{\mathrm{p}2}}{\partial \sigma_{22}} & \dfrac{\partial f_{\mathrm{p}2}}{\partial \sigma_{33}} & \dfrac{\partial f_{\mathrm{p}2}}{\partial \sigma_{12}} & \dfrac{\partial f_{\mathrm{p}2}}{\partial \sigma_{23}} & \dfrac{\partial f_{\mathrm{p}2}}{\partial \sigma_{31}} \\ \cdots & \cdots & \cdots & \cdots & \cdots & \cdots \\ \dfrac{\partial f_{\mathrm{p}m_{\mathrm{p}}}}{\partial \sigma_{11}} & \dfrac{\partial f_{\mathrm{p}m_{\mathrm{p}}}}{\partial \sigma_{22}} & \dfrac{\partial f_{\mathrm{p}m_{\mathrm{p}}}}{\partial \sigma_{33}} & \dfrac{\partial f_{\mathrm{p}m_{\mathrm{p}}}}{\partial \sigma_{12}} & \dfrac{\partial f_{\mathrm{p}m_{\mathrm{p}}}}{\partial \sigma_{23}} & \dfrac{\partial f_{\mathrm{p}m_{\mathrm{p}}}}{\partial \sigma_{31}} \end{bmatrix}_{i-1}. \tag{4.23c}$$

The explicit form of the plastic yield matrix related to the Tresca and von Mises criteria will be presented in the next section.

Finally, we recall that eqns (4.18–4.20) constitute one system of supplementary equations, which will be coupled with eqns (4.4, 4.5) to solve the proposed elastoplastic frictional contact problem. For numerical purposes, eqns (4.18, 4.19) should be assembled for two contacting bodies as follows:

$$\begin{Bmatrix} \{f_{\mathrm{p}}\}_i^1 \\ \{f_{\mathrm{p}}\}_i^2 \end{Bmatrix} = \left\{ \begin{Bmatrix} \{-Y\}^1 \\ \{-Y\}^2 \end{Bmatrix} + \begin{bmatrix} [C_{\mathrm{p}}]_{i-1}^1 & 0 \\ 0 & [C_{\mathrm{p}}]_{i-1}^2 \end{bmatrix} \begin{Bmatrix} \{\sigma\}_{i-1}^1 \\ \{\sigma\}_{i-1}^2 \end{Bmatrix} \right\} + \begin{bmatrix} [C_{\mathrm{p}}]_{i-1}^1 & 0 \\ 0 & [C_{\mathrm{p}}]_{i-1}^2 \end{bmatrix} \begin{Bmatrix} \{\mathrm{d}\sigma\}_i^1 \\ \{\mathrm{d}\sigma\}_i^2 \end{Bmatrix}, \tag{4.24}$$

$$\begin{Bmatrix} \{\mathrm{d}\varepsilon^{\mathrm{p}}\}_i^1 \\ \{\mathrm{d}\varepsilon^{\mathrm{p}}\}_i^2 \end{Bmatrix} = \begin{bmatrix} [[C_{\mathrm{p}}]_{i-1}^{\mathrm{T}}]^1 & 0 \\ 0 & [[C_{\mathrm{p}}]_{i-1}^{\mathrm{T}}]^2 \end{bmatrix} \begin{Bmatrix} \{\mathrm{d}\lambda_{\mathrm{p}}\}_i^1 \\ \{\mathrm{d}\lambda_{\mathrm{p}}\}_i^2 \end{Bmatrix}. \tag{4.25}$$

It is clear that eqn (4.20) is valid for the assemblies.

4.4 General form of the plastic yield matrix

In the present study, the elastoplastic behavior of materials is assumed to follow the Tresca or von Mises criteria. It is well known that most metallic engineering materials, e.g. steel and aluminum, have yield properties that fall somewhere between these two yield criteria. In the previous section, the piecewise linear yield functions approximation was introduced to represent the plastic yield function f_p in the principal stress state with respect to these yield conditions. On the other hand, the term $\partial f_{pr}/\partial\sigma_{ij}$, where $r = 1, \ldots, m_p$ and $i, j = 1, 2, 3$, is the general term of the plastic yield matrix $[C_p]_{i-1}$ as shown in eqns (4.22c) in 2D and (4.23c) in 3D. This term is the derivative of the plastic yield function f_p in the principal stress space, with respect to the stress components. Therefore, it may be alternatively represented:

$$\frac{\partial f_{pr}}{\partial \sigma_{ij}} = \frac{\partial f_{pr}}{\partial \sigma_k} \cdot \frac{\partial \sigma_k}{\partial \sigma_{ij}}, \tag{4.26}$$

where σ_{ij} $(i, j = 1, 2, 3)$ and $\sigma_k(k = 1, 2, 3)$ are the stress and principal stress components respectively. This equation can now be written in matrix form:

$$[C_p]_{i-1} = [C_p]_{\sigma_k} \cdot [[\nabla\sigma_k]_{\sigma_{ij}}]_{i-1}. \tag{4.27}$$

The first right-hand term in the above equation can be found using the Tresca and the von Mises yield conditions. The second one, which is the derivative of the principal stress components with respect to the stress components, can be obtained in a unified form, using a general approach to evaluate the principal stresses presented in Appendix C. Both terms are derived in what follows.

A. Evaluate matrix $[C_p]_{\sigma_k}$

A.1 Tresca yield matrix

The Tresca yield criterion corresponds to a material that yields when the maximum shear stress reaches the value of the maximum shear stress at yielding in the tensile test. Mathematically, yielding occurs when:

$$\text{Max}[|\sigma_1 - \sigma_2|, |\sigma_2 - \sigma_3|, |\sigma_3 - \sigma_1|] = Y. \tag{4.28}$$

Principal stresses are often employed as a convenient means to express this yield criterion. In terms of the principal stresses, the maximum shear stress is given by half the maximum difference between any two of them. In general, the order of magnitude of the principal stresses is not considered and all the possible shear outcomes are embodied in the yield criterion. The single function defining this yield criterion is thus given by:

$$f_p = ((\sigma_1 - \sigma_2)^2 - Y^2)((\sigma_2 - \sigma_3)^2 - Y^2)((\sigma_3 - \sigma_1)^2 - Y^2) = 0, \tag{4.29}$$

in which Y is the initial yield stress. This yield condition can be represented graphically by a regular hexagonal cylinder with its centroidal axis (hydrostatic axis) oriented equidistant from the three principal stress axes $\sigma_k(k = 1, 2, 3)$, as shown in fig. 4.1 in 3D and in fig. 4.2 in 2D (plane stress).

It is convenient to expand eqn (4.29) to a set of six piecewise linear functions, $(m_p = 6)$, each defining one face of the hexagonal yield surface:

$$\begin{Bmatrix} f_{p1} \\ f_{p2} \\ f_{p3} \\ f_{p4} \\ f_{p5} \\ f_{p6} \end{Bmatrix} = \begin{Bmatrix} \sigma_1 - \sigma_3 - Y \\ -\sigma_2 + \sigma_3 - Y \\ \sigma_1 - \sigma_2 - Y \\ \sigma_2 - \sigma_3 - Y \\ -\sigma_1 + \sigma_3 - Y \\ -\sigma_1 + \sigma_2 - Y \end{Bmatrix}. \tag{4.30}$$

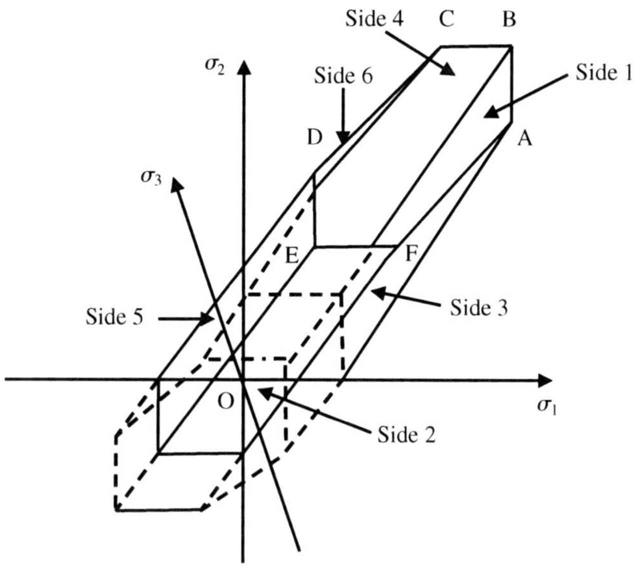

Figure 4.1: 3D Tresca yield condition.

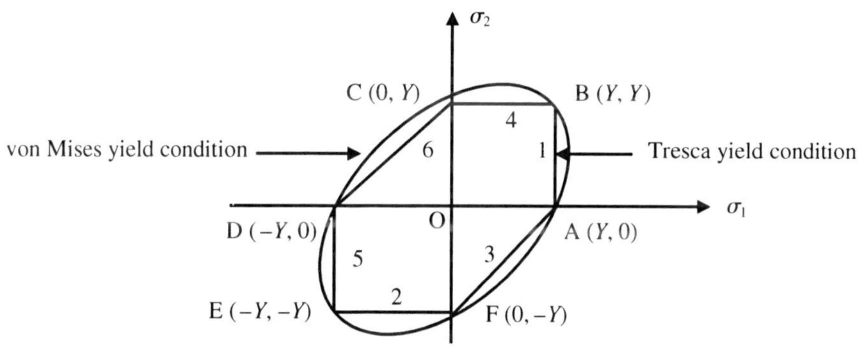

Figure 4.2: Tresca and von Mises yield conditions in plane stress (2D).

Finally, the Tresca yield matrix can now be expressed:

(a) In plane strain and in 3D:

$$[C_\mathrm{p}]_{\sigma_k} = \begin{bmatrix} \dfrac{\partial f_{\mathrm{p}1}}{\partial \sigma_1} & \dfrac{\partial f_{\mathrm{p}1}}{\partial \sigma_2} & \dfrac{\partial f_{\mathrm{p}1}}{\partial \sigma_3} \\ \dfrac{\partial f_{\mathrm{p}2}}{\partial \sigma_1} & \dfrac{\partial f_{\mathrm{p}2}}{\partial \sigma_2} & \dfrac{\partial f_{\mathrm{p}2}}{\partial \sigma_3} \\ \cdots & \cdots & \cdots \\ \dfrac{\partial f_{\mathrm{p}6}}{\partial \sigma_1} & \dfrac{\partial f_{\mathrm{p}6}}{\partial \sigma_2} & \dfrac{\partial f_{\mathrm{p}6}}{\partial \sigma_3} \end{bmatrix}_{\sigma_k} = \begin{bmatrix} 1 & 0 & -1 \\ 0 & -1 & 1 \\ 1 & -1 & 0 \\ 0 & 1 & -1 \\ -1 & 0 & 1 \\ -1 & 1 & 0 \end{bmatrix}_{\sigma_k} . \tag{4.31}$$

(b) In plane stress (2D):

$$[C_\mathrm{p}]_{\sigma_k} = \begin{bmatrix} \frac{\partial f_{\mathrm{p}1}}{\partial \sigma_1} & \frac{\partial f_{\mathrm{p}1}}{\partial \sigma_2} \\ \frac{\partial f_{\mathrm{p}2}}{\partial \sigma_1} & \frac{\partial f_{\mathrm{p}2}}{\partial \sigma_2} \\ \cdots & \cdots \\ \frac{\partial f_{\mathrm{p}6}}{\partial \sigma_1} & \frac{\partial f_{\mathrm{p}6}}{\partial \sigma_2} \end{bmatrix}_{\sigma_k} = \begin{bmatrix} 1 & 0 \\ 0 & -1 \\ 1 & -1 \\ 0 & 1 \\ -1 & 0 \\ -1 & 1 \end{bmatrix}_{\sigma_k}. \tag{4.32}$$

A.2 Von Mises yield matrix

The von Mises initial yield criterion is represented by a smooth circular surface expressed by the relation:

$$f_\mathrm{p} = (\sigma_1 - \sigma_2)^2 + (\sigma_2 - \sigma_3)^2 + (\sigma_3 - \sigma_1)^2 - 2Y^2 = 0. \tag{4.33}$$

For a material obeying the von Mises criterion, yielding occurs when the octahedral shearing stress is equal to that attained at yielding $[(\sqrt{2}/3)Y]$ in a tensile test. This criterion can be represented graphically by a circular cylinder instead of a hexagonal cylinder for the Tresca yield condition as shown in fig. 4.2 for 2D (plane stress) and in fig. 4.3 for 3D. In order to express the von Mises condition as a set of piecewise linear functions, each defining one face of the circular cylinder yield surface (eqn (4.33)) can be given by:

$$f_{\mathrm{p}r} = [\cos\gamma_r \quad \sin\gamma_r \quad 0][N_\sigma]^\mathrm{T} \begin{Bmatrix} \sigma_1 \\ \sigma_2 \\ \sigma_3 \end{Bmatrix} - \frac{\sqrt{2+\sqrt{2}}}{2}\mathrm{Y}, \quad r = 1, \ldots, m_\mathrm{p}, \tag{4.34}$$

where $\gamma_r = (r-1)(2\pi/m_\mathrm{p})$ represents geometrically the normal direction to the sides of the polygon as shown in fig. 4.4, where m_p is assumed to be equal to 8 in the present study. $[N_\sigma]$ is the matrix transforming $\{\sigma_1\ \sigma_2\ \sigma_3\}^\mathrm{T}$ into the $\{\xi\ \eta\ \zeta\}^\mathrm{T}$ space, in which $(\xi\ \eta\ \zeta)$ is a local frame,

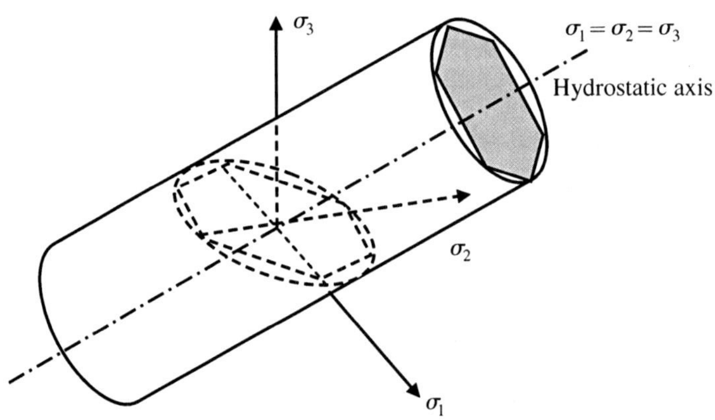

Figure 4.3: 3D Tresca and von Mises yield conditions.

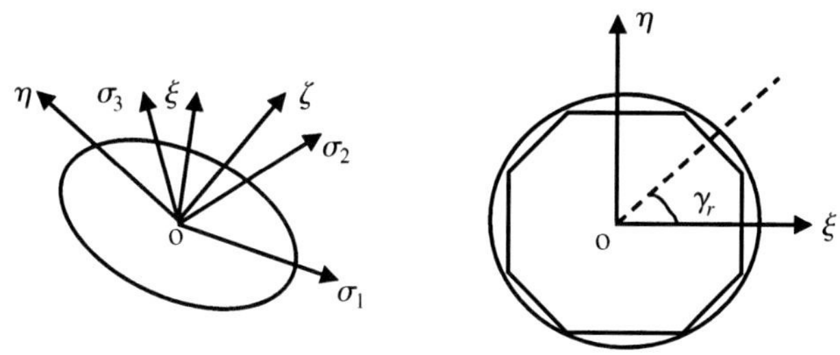

Figure 4.4: Von Mises yield surface and its piecewise linear approximation.

with ζ being the axis of the von Mises cylinder and (ξ η) axes defining the π-plane through the origin, as shown in fig. 4.4 (Kong *et al.*, 1993). The matrix $[N_\sigma]$ can be written:

$$[N_\sigma] = \begin{bmatrix} -\frac{1}{\sqrt{2}} & -\frac{1}{\sqrt{6}} & \frac{1}{\sqrt{3}} \\ \frac{1}{\sqrt{2}} & -\frac{1}{\sqrt{6}} & \frac{1}{\sqrt{3}} \\ 0 & \frac{2}{\sqrt{6}} & \frac{1}{\sqrt{3}} \end{bmatrix}. \tag{4.35}$$

Finally, the von Mises yield matrix can now be expressed:

$$[C_\mathrm{p}]_{\sigma_k} = \begin{bmatrix} \frac{\partial f_{\mathrm{p}1}}{\partial \sigma_1} & \frac{\partial f_{\mathrm{p}1}}{\partial \sigma_2} & \frac{\partial f_{\mathrm{p}1}}{\partial \sigma_3} \\ \frac{\partial f_{\mathrm{p}2}}{\partial \sigma_1} & \frac{\partial f_{\mathrm{p}2}}{\partial \sigma_2} & \frac{\partial f_{\mathrm{p}2}}{\partial \sigma_3} \\ \cdots & \cdots & \cdots \\ \frac{\partial f_{\mathrm{p}8}}{\partial \sigma_1} & \frac{\partial f_{\mathrm{p}8}}{\partial \sigma_2} & \frac{\partial f_{\mathrm{p}8}}{\partial \sigma_3} \end{bmatrix}_{\sigma_k} = \begin{bmatrix} -\frac{1}{\sqrt{2}} & \frac{1}{\sqrt{2}} & 0 \\ -\frac{1}{2} - \frac{1}{2\sqrt{3}} & \frac{1}{2} - \frac{1}{2\sqrt{3}} & \frac{1}{\sqrt{3}} \\ -\frac{1}{\sqrt{6}} & -\frac{1}{\sqrt{6}} & \frac{2}{\sqrt{6}} \\ \frac{1}{2} - \frac{1}{2\sqrt{3}} & -\frac{1}{2} - \frac{1}{2\sqrt{3}} & \frac{1}{\sqrt{3}} \\ \frac{1}{\sqrt{2}} & -\frac{1}{\sqrt{2}} & 0 \\ \frac{1}{2} + \frac{1}{2\sqrt{3}} & -\frac{1}{2} + \frac{1}{2\sqrt{3}} & -\frac{1}{\sqrt{3}} \\ \frac{1}{\sqrt{6}} & \frac{1}{\sqrt{6}} & -\frac{2}{\sqrt{6}} \\ -\frac{1}{2} + \frac{1}{2\sqrt{3}} & \frac{1}{2} + \frac{1}{2\sqrt{3}} & -\frac{1}{\sqrt{3}} \end{bmatrix}_{\sigma_k}. \tag{4.36}$$

Notice that in the plane stress condition ($\sigma_3 = 0$), the above matrix reduces to the first two columns and rows.

It should be noted that a particular case of the matrix $[C_\mathrm{p}]_{\sigma_k}$ may be found in a paper by Kong *et al.* (1993).

B. Evaluate matrix $[\nabla\sigma_k]_{\sigma_{ij}}$

Using the detailed results given in Appendix C and assuming that the principal stresses of the symmetric stress tensor $[\sigma]$ are $\{\sigma_1\ \sigma_2\ \sigma_3\}^\mathrm{T}$ with $\sigma_1 \geq \sigma_2 \geq \sigma_3$, it is then possible to write:

$$\begin{Bmatrix} \sigma_1 \\ \sigma_2 \\ \sigma_3 \end{Bmatrix} = 2\sqrt{\frac{J_2}{3}} \begin{Bmatrix} \cos\omega_1 \\ \cos\omega_2 \\ \cos\omega_3 \end{Bmatrix} + \frac{I_1}{3} \begin{Bmatrix} 1 \\ 1 \\ 1 \end{Bmatrix}, \tag{4.37}$$

where

$$\omega_1 = \frac{1}{3}\left[\cos^{-1}\left(\frac{J_3}{2}\left(\frac{3}{J_2}\right)^{3/2}\right)\right], \quad \langle\omega_2 \quad \omega_3\rangle = \left\langle\omega_1 - \frac{2\pi}{3} \quad \omega_1 + \frac{2\pi}{3}\right\rangle, \quad 0 \le \omega_1 \le \pi/3. \tag{4.38}$$

(a) In 2D (in plane stress: $\sigma_3 = \sigma_{33} = 0$):

$$\begin{aligned} I_1 &= \sigma_{11} + \sigma_{22} + \sigma_{33}, \\ J_2 &= \frac{1}{2}(\sigma_{11}'^2 + \sigma_{22}'^2 + \sigma_{33}'^2) + \sigma_{12}^2, \\ J_3 &= \sigma_{33}'(\sigma_{33}'^2 - J_2). \end{aligned} \tag{4.39a}$$

(b) In 3D:

$$\begin{aligned} I_1 &= \sigma_{11} + \sigma_{22} + \sigma_{33}, \\ J_2 &= \frac{1}{2}(\sigma_{11}'^2 + \sigma_{22}'^2 + \sigma_{33}'^2) + \sigma_{12}^2 + \sigma_{23}^2 + \sigma_{31}^2, \\ J_3 &= \sigma_{11}'\sigma_{22}'\sigma_{33}' + 2\sigma_{12}\sigma_{23}\sigma_{31} - (\sigma_{11}'\sigma_{23}^2 + \sigma_{22}'\sigma_{31}^2 + \sigma_{33}'\sigma_{12}^2) \end{aligned} \tag{4.39b}$$

in which the deviatoric components are:

$$\begin{aligned} \sigma_{11}' &= \sigma_{11} - \frac{I_1}{3}, \\ \sigma_{22}' &= \sigma_{22} - \frac{I_1}{3}, \\ \sigma_{33}' &= \sigma_{33} - \frac{I_1}{3}. \end{aligned} \tag{4.40}$$

The matrix $[\nabla\sigma_k]_{\sigma_{ij}}$, which is the second term of eqn (4.27), may be explicitly written:

(a) In 2D:

$$\begin{bmatrix} \dfrac{\partial\sigma_1}{\partial\sigma_{11}} & \dfrac{\partial\sigma_1}{\partial\sigma_{22}} & \dfrac{\partial\sigma_1}{\partial\sigma_{12}} \\ \dfrac{\partial\sigma_2}{\partial\sigma_{11}} & \dfrac{\partial\sigma_2}{\partial\sigma_{22}} & \dfrac{\partial\sigma_2}{\partial\sigma_{12}} \\ \dfrac{\partial\sigma_3}{\partial\sigma_{11}} & \dfrac{\partial\sigma_3}{\partial\sigma_{22}} & \dfrac{\partial\sigma_3}{\partial\sigma_{12}} \end{bmatrix} = \begin{bmatrix} (D_2)_1 & (D_3)_1 & (D_1)_1 \\ (D_2)_2 & (D_3)_2 & (D_1)_2 \\ (D_2)_3 & (D_3)_3 & (D_1)_3 \end{bmatrix} \cdot \begin{bmatrix} \dfrac{\partial J_2}{\partial\sigma_{11}} & \dfrac{\partial J_2}{\partial\sigma_{22}} & \dfrac{\partial J_2}{\partial\sigma_{12}} \\ \dfrac{\partial J_3}{\partial\sigma_{11}} & \dfrac{\partial J_3}{\partial\sigma_{22}} & \dfrac{\partial J_3}{\partial\sigma_{12}} \\ \dfrac{\partial I_1}{\partial\sigma_{11}} & \dfrac{\partial I_1}{\partial\sigma_{22}} & \dfrac{\partial I_1}{\partial\sigma_{12}} \end{bmatrix}, \tag{4.41a}$$

$$\left\langle\frac{\partial J_2}{\partial\sigma_{11}} \quad \frac{\partial J_2}{\partial\sigma_{22}} \quad \frac{\partial J_2}{\partial\sigma_{12}}\right\rangle = \langle\sigma_{11}' \quad \sigma_{22}' \quad 2\sigma_{12}\rangle, \tag{4.41b}$$

$$\left\langle\frac{\partial J_3}{\partial\sigma_{11}} \quad \frac{\partial J_3}{\partial\sigma_{22}} \quad \frac{\partial J_3}{\partial\sigma_{12}}\right\rangle = \langle\sigma_{22}'\sigma_{33}' \quad \sigma_{33}'\sigma_{11}' \quad 2\sigma_{33}'\sigma_{12}\rangle + \frac{1}{3}J_2\,\langle 1 \quad 1 \quad 0\rangle, \tag{4.41c}$$

$$\left\langle\frac{\partial I_1}{\partial\sigma_{11}} \quad \frac{\partial I_1}{\partial\sigma_{22}} \quad \frac{\partial I_1}{\partial\sigma_{12}}\right\rangle = \langle 1 \quad 1 \quad 0\rangle. \tag{4.41d}$$

(b) In 3D:

$$\begin{bmatrix} \frac{\partial\sigma_1}{\partial\sigma_{11}} & \frac{\partial\sigma_1}{\partial\sigma_{22}} & \frac{\partial\sigma_1}{\partial\sigma_{33}} & \frac{\partial\sigma_1}{\partial\sigma_{12}} & \frac{\partial\sigma_1}{\partial\sigma_{23}} & \frac{\partial\sigma_1}{\partial\sigma_{31}} \\ \frac{\partial\sigma_2}{\partial\sigma_{11}} & \frac{\partial\sigma_2}{\partial\sigma_{22}} & \frac{\partial\sigma_2}{\partial\sigma_{33}} & \frac{\partial\sigma_2}{\partial\sigma_{12}} & \frac{\partial\sigma_2}{\partial\sigma_{23}} & \frac{\partial\sigma_2}{\partial\sigma_{31}} \\ \frac{\partial\sigma_3}{\partial\sigma_{11}} & \frac{\partial\sigma_3}{\partial\sigma_{22}} & \frac{\partial\sigma_3}{\partial\sigma_{33}} & \frac{\partial\sigma_3}{\partial\sigma_{12}} & \frac{\partial\sigma_3}{\partial\sigma_{23}} & \frac{\partial\sigma_3}{\partial\sigma_{31}} \end{bmatrix}$$

$$= \begin{bmatrix} (D_2)_1 & (D_3)_1 & (D_1)_1 \\ (D_2)_2 & (D_3)_2 & (D_1)_2 \\ (D_2)_3 & (D_3)_3 & (D_1)_3 \end{bmatrix} \cdot \begin{bmatrix} \frac{\partial J_2}{\partial\sigma_{11}} & \frac{\partial J_2}{\partial\sigma_{22}} & \frac{\partial J_2}{\partial\sigma_{33}} & \frac{\partial J_2}{\partial\sigma_{12}} & \frac{\partial J_2}{\partial\sigma_{23}} & \frac{\partial J_2}{\partial\sigma_{31}} \\ \frac{\partial J_3}{\partial\sigma_{11}} & \frac{\partial J_3}{\partial\sigma_{22}} & \frac{\partial J_3}{\partial\sigma_{33}} & \frac{\partial J_3}{\partial\sigma_{12}} & \frac{\partial J_3}{\partial\sigma_{23}} & \frac{\partial J_3}{\partial\sigma_{31}} \\ \frac{\partial I_1}{\partial\sigma_{11}} & \frac{\partial I_1}{\partial\sigma_{22}} & \frac{\partial I_1}{\partial\sigma_{33}} & \frac{\partial I_1}{\partial\sigma_{12}} & \frac{\partial I_1}{\partial\sigma_{23}} & \frac{\partial I_1}{\partial\sigma_{31}} \end{bmatrix}, \tag{4.42a}$$

$$\left\langle \frac{\partial J_2}{\partial\sigma_{11}} \quad \frac{\partial J_2}{\partial\sigma_{22}} \quad \frac{\partial J_2}{\partial\sigma_{33}} \quad \frac{\partial J_2}{\partial\sigma_{12}} \quad \frac{\partial J_2}{\partial\sigma_{23}} \quad \frac{\partial J_2}{\partial\sigma_{31}} \right\rangle = \left\langle \sigma'_{11} \quad \sigma'_{22} \quad \sigma'_{33} \quad 2\sigma_{12} \quad 2\sigma_{23} \quad 2\sigma_{31} \right\rangle, \tag{4.42b}$$

$$\left\langle \frac{\partial J_3}{\partial\sigma_{11}} \quad \frac{\partial J_3}{\partial\sigma_{22}} \quad \frac{\partial J_3}{\partial\sigma_{33}} \quad \frac{\partial J_3}{\partial\sigma_{12}} \quad \frac{\partial J_3}{\partial\sigma_{23}} \quad \frac{\partial J_3}{\partial\sigma_{31}} \right\rangle = \left\langle \begin{matrix} \sigma'_{22}\sigma'_{33} - \sigma_{23}^2 \quad \sigma'_{33}\sigma'_{11} - \sigma_{31}^2 \quad \sigma'_{11}\sigma'_{22} - \sigma_{12}^2 \\ 2(\sigma_{23}\sigma_{31} - \sigma'_{33}\sigma_{12}) \quad 2(\sigma_{31}\sigma_{12} - \sigma'_{11}\sigma_{23}) \quad 2(\sigma_{12}\sigma_{23} - \sigma'_{22}\sigma_{31}) \end{matrix} \right\rangle + \frac{1}{3} J_2 \langle 1 \quad 1 \quad 1 \quad 0 \quad 0 \quad 0 \rangle, \tag{4.42c}$$

$$\left\langle \frac{\partial I_1}{\partial\sigma_{11}} \quad \frac{\partial I_1}{\partial\sigma_{22}} \quad \frac{\partial I_1}{\partial\sigma_{33}} \quad \frac{\partial I_1}{\partial\sigma_{12}} \quad \frac{\partial I_1}{\partial\sigma_{23}} \quad \frac{\partial I_1}{\partial\sigma_{31}} \right\rangle = \langle 1 \quad 1 \quad 1 \quad 0 \quad 0 \quad 0 \rangle, \tag{4.42d}$$

where

$$\langle (D_3)_1 \quad (D_3)_2 \quad (D_3)_3 \rangle = \frac{1}{J_2} \cdot \frac{1}{\sin 3\omega_1} \langle \sin\omega_1 \quad \sin\omega_2 \quad \sin\omega_3 \rangle,$$

$$\langle (D_2)_1 \quad (D_2)_2 \quad (D_2)_3 \rangle = \left\langle \frac{\cos\omega_1}{\sqrt{3J_2}} - \left(\frac{3J_3}{2J_2}\right)(D_3)_1 \quad \frac{\cos\omega_2}{\sqrt{3J_2}} - \left(\frac{3J_3}{2J_2}\right)(D_3)_2 \quad \frac{\cos\omega_3}{\sqrt{3J_2}} - \left(\frac{3J_3}{2J_2}\right)(D_3)_3 \right\rangle,$$

$$\langle (D_1)_1 \quad (D_1)_2 \quad (D_1)_3 \rangle = \frac{1}{3} \langle 1 \quad 1 \quad 1 \rangle. \tag{4.43}$$

Numerical difficulties are encountered as ω_1 approaches 0 and $\pi/3$, situations that arise at the corner of the Tresca and linearized von Mises criteria as can be seen from eqn (4.43), since for these values of ω_1 both $(D_3)_{1,2,3}$ and $(D_2)_{1,2,3}$ become indeterminate. On the other hand, values of $\omega_1 = 0$ and $\pi/3$ correspond to singular points on the yield surface where the direction of plastic straining is indeterminate. This difficulty can be overcome by returning to the original eqn (4.37) and rewriting this for the explicit values when $\omega_1 = 0$ and $\pi/3$, giving:

$$\begin{Bmatrix} \sigma_1 \\ \sigma_2 \\ \sigma_3 \end{Bmatrix} = 2\sqrt{\frac{J_2}{3}} \begin{Bmatrix} V_1 \\ V_2 \\ V_3 \end{Bmatrix} + \frac{I_1}{3} \begin{Bmatrix} 1 \\ 1 \\ 1 \end{Bmatrix}, \tag{4.44}$$

where

$$\begin{aligned} \omega_1 = 0 &\rightarrow \langle V_1 \quad V_2 \quad V_3 \rangle = \left\langle 1 \quad -\frac{1}{2} \quad -\frac{1}{2} \right\rangle, \\ \omega_2 = \frac{\pi}{3} &\rightarrow \langle V_1 \quad V_2 \quad V_3 \rangle = \left\langle \frac{1}{2} \quad \frac{1}{2} \quad -1 \right\rangle. \end{aligned} \tag{4.45}$$

The practical approach adopted in the present study is to use the general eqn (4.43) for both $(D_3)_{1,2,3}$ and $(D_2)_{1,2,3}$ for all values between $1^\circ \le \omega_1 \le 59^\circ$. Outside this interval, we employ the following relation, which makes the direction of straining unique and physically corresponds to a rounding off of the yield surface corners.

$$\begin{aligned} \langle (D_3)_1 \quad (D_3)_2 \quad (D_3)_3 \rangle &= \langle 0 \quad 0 \quad 0 \rangle, \\ \langle (D_2)_1 \quad (D_2)_2 \quad (D_2)_3 \rangle &= \frac{1}{\sqrt{3J_2}} \langle V_1 \quad V_2 \quad V_3 \rangle, \\ \langle (D_1)_1 \quad (D_1)_2 \quad (D_1)_3 \rangle &= \frac{1}{3} \langle 1 \quad 1 \quad 1 \rangle. \end{aligned} \tag{4.46}$$

It should be noted that an alternative form of matrix $[\nabla \sigma_k]_{\sigma_{ij}}$ can be found in a paper by Nayak and Zienkiewicz (1972), as well as in the book of Owen and Hinton (1980).

4.5 Boundary element formulation for elastoplastic frictional contact problems

Frictional contact problems are highly nonlinear not only due to the unknown contact area but also due to the unknown contact conditions when frictional phenomena are present. By including another nonlinearity due to elastoplastic material behavior, to take into account important cases where the stress concentrations in the contact surface are serious enough to produce plastic deformation, the problem becomes increasingly complicated.

Since the contact boundary conditions usually occur on the surface or boundary of two contacting bodies and since more than one body must always be discretized in contact problems, the boundary element method (BEM), which discretizes only the boundaries of the contacting bodies and in the case of local plasticity, the region where plastic deformation is expected, is a powerful method to analyze contact problems between elastoplastic bodies. Although elastoplastic contact analyses using the BEM require a domain discretization, using the particular integrals formulation (meshless method) introduced by Henry and Banerjee (1988) can completely eliminate the need for volume integrals. Therefore, the major significant advantage of the BEM over the FEM is that only a surface discretization is maintained. This is the principal reason why the BEM provides such an efficient and suitable numerical discretization method for the analysis of elastoplastic contact problems.

The use of the BEM in the present study provides the necessary and sufficient relations between the contact variables as well as the stress field, in incremental form for treating contacting bodies.

It supplies the last supplementary equations required to solve the proposed elastoplastic frictional contact problem presented by eqns (4.4, 4.5), (4.20) and (4.24, 4.25).

In order to derive the boundary element formulation for elastoplastic frictional contact problems, we start by presenting the boundary integral equations for each body. The presentation is based on a particular integral concept and follows the notation proposed by Henry and Banerjee (1988). Let us consider one of the isotropic elastoplastic contacting bodies before contact takes place (fig. 2.1). Recalling the set of standard governing equations presented in Chapter 2, one can rewrite them in incremental form.

(1) Static equilibrium equations:

$$\mathrm{d}\sigma_{ij,i} = 0 \quad \text{in } V \text{ (without body force)} \quad \text{and} \quad i, j = 1, 2, 3, \tag{4.47}$$

where $\mathrm{d}\sigma_{ij}$ are the incremental stress components.

(2) Boundary conditions:

(b1) Essential conditions or prescribed displacements:

$$\mathrm{d}u_i = \mathrm{d}\bar{u}_i \quad \text{on } S_u \quad i = 1, 2, 3. \tag{4.48a}$$

(b2) Natural conditions or prescribed tractions:

$$\mathrm{d}t_i = \Gamma \mathrm{d}\bar{t}_i \quad \text{on } S_f \quad i = 1, 2, 3, \tag{4.48b}$$

where it is assumed that the surface tractions $\mathrm{d}\bar{t}$ (the external loads) increase proportionally with the load factor Γ.

(3) Displacement–strain relationships (compatibility or kinematic equations):

$$\mathrm{d}\varepsilon_{ij} = \frac{1}{2}(\mathrm{d}u_{i,j} + \mathrm{d}u_{j,i}) = \mathrm{d}\varepsilon^{\mathrm{e}}_{ij} + \mathrm{d}\varepsilon^{\mathrm{o}}_{ij} \quad \text{in } V \text{ and } i, j = 1, 2, 3, \tag{4.49}$$

where the incremental total strain, $\mathrm{d}\varepsilon_{ij}$, determined from incremental displacement spatial derivatives, $\mathrm{d}u_{i,j}$, has been partitioned into a recoverable incremental elastic component, $\mathrm{d}\varepsilon^{\mathrm{e}}_{ij}$, and a permanent incremental inelastic (plastic) component $\mathrm{d}\varepsilon^{\mathrm{o}}_{ij}$. It should be noted that $\mathrm{d}\varepsilon^{\mathrm{o}}_{ij}$ can include any kind of inelastic strain such as plastic, creep, thermal or others, although the work carried out here deals only with inelastic strains caused by plasticity. The permanent incremental plastic (inelastic) component is often referred to as the incremental initial strain component. In this chapter, the superscript ‘o’ will be used to indicate these inelastic quantities.

(4) Strain–stress relationships (material constitutive law or generalized Hooke’s law):

$$\mathrm{d}\sigma_{ij} = \frac{E}{(1+\nu)}\left[\frac{\nu}{(1-2\nu)}\delta_{ij}(\mathrm{d}\varepsilon_{kk} - \mathrm{d}\varepsilon) + (\mathrm{d}\varepsilon_{ij} - \mathrm{d}\varepsilon^{\mathrm{o}}_{ij})\right] \quad \text{in } V \text{ and } i, j = 1, 2, 3, \tag{4.50}$$

where E and ν are Young’s modulus and Poisson’s ratio respectively, and δ_{ij} is the Kronecker delta symbol ($\equiv 1$ for $i = j$ and $\equiv 0$ for $i \neq j$). Furthermore, the incremental strain, $\mathrm{d}\varepsilon_{kk}$, can be expressed as the total volumetric strain:

$$\mathrm{d}\varepsilon_{kk} = \mathrm{d}\varepsilon_{11} + \mathrm{d}\varepsilon_{22} + \mathrm{d}\varepsilon_{33}. \tag{4.51}$$

Notice that $\mathrm{d}\varepsilon = \mathrm{d}\varepsilon^{\mathrm{o}}_{kk}$, i.e. the volumetric incremental component of the plastic (inelastic) dilatational strain is zero since the plastic deformation is incompressible.

It is interesting to rewrite the strain–stress relationships by separating the contributions in the stress computation into those due to the incremental total strain and those due to the so-called initial incremental stress, $\mathrm{d}\sigma_{ij}^{\mathrm{o}}$:

$$\begin{aligned}\mathrm{d}\sigma_{ij} &= \frac{E}{(1+\nu)}\left[\frac{\nu}{(1-2\nu)}\delta_{ij}\mathrm{d}\varepsilon_{kk} + \mathrm{d}\varepsilon_{ij}\right] - \mathrm{d}\sigma_{ij}^{\mathrm{o}} \\ &= D_{ijkl}^{\mathrm{e}}\mathrm{d}\varepsilon_{kl} - \mathrm{d}\sigma_{ij}^{\mathrm{o}},\end{aligned} \tag{4.52}$$

where D_{ijkl}^{e} is the elastic constitutive tensor given by:

$$D_{ijkl}^{\mathrm{e}} = \frac{E}{(1+\nu)}\left[\frac{\nu}{(1-2\nu)}\delta_{ij}\delta_{kl} + \delta_{ik}\delta_{jl}\right]. \tag{4.53}$$

As in the theory of elasticity (Chapter 2), the governing differential equations for elastoplastic behavior, also called Navier's incremental equations, can be obtained by differentiating eqn (4.52) and substituting it into the stress equilibrium relation (eqn (4.47)) using eqn (4.49), and can be written in terms of incremental displacements:

$$\frac{E}{2(1+\nu)}\mathrm{d}u_{i,jj} + \frac{E}{2(1+\nu)(1-2\nu)}\mathrm{d}u_{j,ji} = \mathrm{d}\sigma_{ij,j}^{\mathrm{o}} \quad \text{in } V \text{ and } i, j = 1, 2, 3. \tag{4.54}$$

These equations are in terms of the incremental initial stresses. However, working with eqn (4.53), they can also be represented in terms of the incremental initial strains:

$$\mathrm{d}u_{i,jj} + \frac{1}{(1-2\nu)}\mathrm{d}u_{j,ji} = 2\left[\frac{\nu}{(1-2\nu)}\mathrm{d}\varepsilon_{,i} + \mathrm{d}\varepsilon_{ij,j}^{\mathrm{o}}\right]. \tag{4.55}$$

Note that the substitution properties of the Kronecker delta have been used to derive eqns (4.54, 4.55).

All the above equations are used both for the three-dimensional analysis and for the plane strain case ($i, j = 1, 2$). When the plane stress condition is assumed ($i, j = 1, 2$), the same formulation is maintained simply by replacing E by $E(1+2\nu)/(1+\nu)^2$ and ν by $\nu/(1+\nu)$. It is well known that, in plane strain, the strain component in the third dimension is zero. However, the stress component in that direction is generally nonzero. On the other hand, in plane stress, the stress component in the third dimension is zero. However, the strain component in that direction is generally nonzero.

In order to derive the boundary integral equations for elastoplastic problems, two main approaches may be considered. The first approach requires a discretization of the domain of the problem into volume cells (many analogous to finite elements), i.e. volume integration is required, thus losing one of the major advantages of the boundary element formulation. Numerical integration of these integrals is time consuming and expensive. Furthermore, the volume integral of the stress (strain) equation exhibits strong singularities, making accurate numerical integration difficult (Henry and Banerjee, 1988). However, since then improvements have been made in the numerical evaluation of hypersingular volume and surface integrals (Guiggiani and Gigante, 1990; Guiggiani *et al.*, 1992). In the second approach, employed in the present study, the need for such volume discretization is removed by the use of the concept of particular integrals (meshless method), which incorporate the nonlinear effects into the boundary integral equations instead of volume integrals. Instead of volume cells, this approach only requires a limited number of interior (domain) sampling points to handle the particular integrals (solutions). The final system equation resulting from this approach is identical in form to that obtained by volume integration. The method is based on the well-known concept of developing a solution of a nonhomogeneous differential equation by means of a complementary function and a particular integral.

Navier's incremental equations for elastoplasticity given in eqn (4.54) are the governing nonhomogeneous differential equations and can be written in operator notation:

$$L(\mathrm{d}u_i) = \mathrm{d}\sigma^{\mathrm{o}}_{ij,j}, \tag{4.56}$$

where $L(\mathrm{d}u_i)$ is a self-adjoint, homogeneous, differential operator.

Recalling the theory of differential equations, it is well known that the solution of the above nonhomogeneous differential equation can be accomplished in three parts. First, a complementary solution (function) $\mathrm{d}u_i^{\mathrm{c}}$ satisfying the following homogeneous differential equation is found:

$$L(\mathrm{d}u_i^{\mathrm{c}}) = 0. \tag{4.57}$$

Next, a particular solution or a particular integral $\mathrm{d}u_i^{\mathrm{p}}$ satisfying the following nonhomogeneous differential equation is obtained:

$$L(\mathrm{d}u_i^{\mathrm{p}}) = d\sigma^{\mathrm{o}}_{ij,j}. \tag{4.58}$$

Finally, the combination of the two solutions yields the incremental total displacement $\mathrm{d}u_i = \mathrm{d}u_i^{\mathrm{c}} + \mathrm{d}u_i^{\mathrm{p}}$, which is then used to match the boundary conditions.

In this procedure, the complementary solution is the elasticity solution in incremental form corresponding to the boundary equation analysis without the nonhomogeneous term, i.e. incremental initial stresses, presented in eqn (4.57). Recalling eqn (2.55) from Chapter 2, the complementary solution for the incremental displacement at the boundary nodes related to the homogeneous equation can be expressed in the matrix form:

$$[H]\{\mathrm{d}u^{\mathrm{c}}\} = [G]\{\mathrm{d}t^{\mathrm{c}}\}, \tag{4.59}$$

where vectors $\{\mathrm{d}u^{\mathrm{c}}\}$ and $\{\mathrm{d}t^{\mathrm{c}}\}$ are the complementary solutions for incremental displacement and traction respectively.

Similarly, using eqn (A.35) from Appendix A, the complementary solution for both incremental surface (boundary) and interior stresses related to the homogeneous equation can be written in matrix form:

$$\{\mathrm{d}\sigma^{\mathrm{c}}\} = [G^{\sigma}]\{\mathrm{d}t^{\mathrm{c}}\} - [H^{\sigma}]\{\mathrm{d}u^{\mathrm{c}}\}. \tag{4.60}$$

The complete solutions of the nonhomogeneous differential equation, composed of a complementary and particular solution for surface (boundary) nodal displacement and traction response, as well as stress response at both surface (boundary) and interior sample points, are collected into column vectors written for incremental total displacement $\{\mathrm{d}u\}$, traction $\{\mathrm{d}t\}$ and stress $\{\mathrm{d}\sigma\}$ respectively:

$$\{\mathrm{d}u\} = \{\mathrm{d}u^{\mathrm{c}}\} + \{\mathrm{d}u^{\mathrm{p}}\}, \tag{4.61}$$

$$\{\mathrm{d}t\} = \{\mathrm{d}t^{\mathrm{c}}\} + \{\mathrm{d}t^{\mathrm{p}}\}, \tag{4.62}$$

$$\{\mathrm{d}\sigma\} = \{\mathrm{d}\sigma^{\mathrm{c}}\} + \{\mathrm{d}\sigma^{\mathrm{p}}\}, \tag{4.63}$$

where $\{\mathrm{d}u^{\mathrm{p}}\}$, $\{\mathrm{d}t^{\mathrm{p}}\}$ and $\{\mathrm{d}\sigma^{\mathrm{p}}\}$ are the particular integrals (solutions) for incremental displacement, traction and stress respectively.

Substitution of eqns (4.61–4.63) into eqns (4.59, 4.60) introduces the particular integrals (solutions) into the complementary solution (elastic solution):

$$[H]\{\mathrm{d}u\} = [G]\{\mathrm{d}t\} - \{[G]\{\mathrm{d}t^{\mathrm{p}}\} - [H]\{\mathrm{d}u^{\mathrm{p}}\}\} \tag{4.64}$$

and

$$\{d\sigma\} = [G^{\sigma}]\{dt\} - [H^{\sigma}]\{du\} - \{[G^{\sigma}]\{dt^{p}\} - [H^{\sigma}]\{du^{p}\} - \{d\sigma^{p}\}\}. \quad (4.65)$$

These equations can be solved for the total response increment, if the particular integrals (solutions) are known. Therefore, we seek to determine this solution, which is a solution to the nonhomogeneous differential equation presented in eqn (4.58). This equation, i.e. Navier's incremental equations, however, consists only of three second-order partial differential equations in three unknown displacements $\{du^{p}\}$ ($d\sigma_{ij}^{o}$ is assumed to be known), but these three equations are coupled in terms of the incremental displacement components. Therefore, one cannot just solve one of the three differential equations at a time. In order to reformulate Navier's incremental equations in the uncoupled form, several approaches such as the method of variation of parameters, the method of undetermined coefficients, the Galerkin vector method, etc. exist, but we chose the Galerkin vector approach in our study.

In the Galerkin vector approach (Kane, 1994), the incremental displacement component du_i^{p} is replaced in eqn (4.58) by an expression involving the second derivatives of a so-called Galerkin vector dg_i :

$$\frac{E}{(1+\nu)} du_i^{p} = c\, dg_{i,jj} - dg_{j,ji}. \quad (4.66)$$

The factor $E/(1+\nu)$ has been written on the left side for notational convenience. It should be noted that the constant 'c' has yet to be determined. Substituting eqn (4.66) into eqn (4.58) yields:

$$\left(\frac{c}{2} dg_{i,kk} - \frac{1}{2} dg_{k,ki}\right)_{,jj} + \frac{1}{(1-2\nu)}\left(\frac{c}{2} dg_{j,kk} - \frac{1}{2} dg_{k,kj}\right)_{,ji} = d\sigma_{ij,j}^{o} \quad (4.67a)$$

or

$$\frac{c}{2} dg_{i,kkjj} + dg_{k,kjji}\left(-\frac{1}{2} + \frac{c}{2(1-2\nu)} - \frac{1}{2(1-2\nu)}\right) = d\sigma_{ij,j}^{o}. \quad (4.67b)$$

Note that in the above derivation, we made use of the fact that '$g_{k,kijj} = g_{j,kkji} = g_{k,kjji}$'. The constant 'c' can now be determined as the constant that makes the quantity in parentheses in eqn (4.67) vanish. That is to say:

$$c = 2(1-\nu) \quad (4.68)$$

and finally the 2- and 3D particular integrals for incremental displacement are related to the Galerkin vector dg_i by:

$$du_i^{p} = \frac{2(1-\nu^2)}{E} dg_{i,jj} - \frac{(1+\nu)}{E} dg_{j,ji}. \quad (4.69)$$

Substituting this equation into eqn (4.58), one obtains a relationship between the Galerkin vector and the incremental initial stress function:

$$(1-\nu) dg_{i,kkjj} = d\sigma_{ij,j}^{o}. \quad (4.70)$$

Notice that the distinctive feature of this equation is that it is completely decoupled. The character of this relation suggests that it is advantageous to write the incremental initial stress function in terms of a second order tensor, rather than a vector, by introducing dh_{ij}, i.e. a tensor function:

$$dh_{ij,mmnn} = d\sigma_{ij}^{o}. \quad (4.71)$$

Substituting eqn (4.71) into eqn (4.70) yields the relation between the Galerkin vector and this new function:

$$(1-\nu)\mathrm{d}g_i = \mathrm{d}h_{ij,j}. \tag{4.72}$$

Finally, substituting this result into eqn (4.69), we obtain the desired particular solution of the nonhomogeneous differential equation:

$$\mathrm{d}u_i^{\mathrm{p}} = \frac{2(1+\nu)}{E}\mathrm{d}h_{ij,jkk} - \frac{(1+\nu)}{E(1-\nu)}\mathrm{d}h_{kj,jki}. \tag{4.73}$$

On the basis of the above equation, one may derive the following relations:

(1) Particular integrals for the incremental strain:

$$\mathrm{d}\varepsilon_{ij}^{\mathrm{p}} = \frac{1}{2}(\mathrm{d}u_{i,j}^{\mathrm{p}} + \mathrm{d}u_{j,i}^{\mathrm{p}}). \tag{4.74}$$

(2) Particular integrals for the incremental stress: eqn (4.52) may be expressed as:

$$\begin{aligned}\mathrm{d}\sigma_{ij} &= D_{ijkl}^{\mathrm{e}}\,\mathrm{d}\varepsilon_{kl} - \mathrm{d}\sigma_{ij}^{\mathrm{o}} = D_{ijkl}^{\mathrm{e}}(\mathrm{d}\varepsilon_{kl}^{\mathrm{c}} + \mathrm{d}\varepsilon_{kl}^{\mathrm{p}}) - \mathrm{d}\sigma_{ij}^{\mathrm{o}}\\ &= D_{ijkl}^{\mathrm{e}}\,\mathrm{d}\varepsilon_{kl}^{\mathrm{c}} + (D_{ijkl}^{\mathrm{e}}\,\mathrm{d}\varepsilon_{kl}^{\mathrm{p}} - \mathrm{d}\sigma_{ij}^{\mathrm{o}}) \equiv \mathrm{d}\sigma_{ij}^{\mathrm{c}} + \mathrm{d}\sigma_{ij}^{\mathrm{p}},\end{aligned} \tag{4.75}$$

then

$$\mathrm{d}\sigma_{ij}^{\mathrm{c}} = D_{ijkl}^{\mathrm{e}}\,\mathrm{d}\varepsilon_{kl}^{\mathrm{c}} \tag{4.76}$$

so that

$$\mathrm{d}\sigma_{ij}^{\mathrm{p}} = D_{ijkl}^{\mathrm{e}}\;\mathrm{d}\varepsilon_{kl}^{\mathrm{p}} - \mathrm{d}\sigma_{ij}^{\mathrm{o}}. \tag{4.77}$$

(3) Particular integrals for the incremental traction:

$$\mathrm{d}t_i^{\mathrm{p}} = \mathrm{d}\sigma_{ij}^{\mathrm{p}}\,n_j. \tag{4.78}$$

As shown in the above equations, the particular integrals (solutions) are directly related to the distribution of the unknown tensor $\mathrm{d}h_{ij}$, i.e. the unknown incremental initial stress function $\mathrm{d}\sigma_{ij}^{\mathrm{o}}$, which will be our task to determine. One successful approach (Henry and Banerjee, 1988) has been to represent it by an infinite series, which, for numerical purposes, can be truncated at a finite number of N terms, as shown below:

$$\mathrm{d}h_{ij}(x) = \sum_{n=1}^{N} C(x,\xi_n)\mathrm{d}\psi_{ij}(\xi_n) \quad i,j = 1,2 \text{ in 2D and } 1,2,3 \text{ in 3D}, \tag{4.79}$$

where

$$C(x,\xi_n) = A_0^4(\rho^4 - b_{\mathrm{n}}\rho^5), \tag{4.80}$$

in which $C(x,\xi_n)$ is a global shape function, A_0 a characteristic length, ρ the Euclidean distance between the field (sample) point x and the source (load) point ξ_n, which can be calculated as $\rho = \sqrt{y_i y_i}$ where $y_i = [x_i - (\xi_n)_i]$, b_{n} a constant that can generally be set equal to unity, and $\mathrm{d}\psi_{ij}(\xi_n)$ a fictitious tensor density. Note that all distances are non-dimensionalized by a characteristic length A_0. In order to obtain an accurate distribution of the tensor $\mathrm{d}h_{ij}$, i.e. the distribution of $\mathrm{d}\sigma_{ij}^{\mathrm{o}}$, in any region, the ξ_n coordinates are generally chosen to be the boundary nodes in the problem, along with a group of auxiliary or fictitious nodes in the domain of the regions in question. It is recommended that the distribution of these auxiliary nodes be roughly consistent with the boundary discretization.

Making use of this expression for $\mathrm{d}h_{ij}$, we are now ready to determine the following formula for the different variables of the particular integrals (solutions). The incremental initial stress function, $\mathrm{d}\sigma_{ij}^{\mathrm{o}}$, can be obtained by substituting four differentiations of $\mathrm{d}h_{ij}$ from eqn (4.79) into eqn (4.71):

$$\mathrm{d}\sigma_{ij}^{\mathrm{o}}(x) = \mathrm{d}h_{ij,mmll}(x) = \sum_{n=1}^{N} K(x, \xi_n)\mathrm{d}\psi_{ij}(\xi_n), \tag{4.81}$$

where

$$K(x, \xi_n) = C_{,mmll}(x, \xi_n) = A_0^4(a - b\rho), \tag{4.82a}$$

$$a = 8d(d+2), \quad b = 15b_{\mathrm{n}}(d+3)(d+1), \quad d = 2 \text{ in 2D and 3 in 3D.} \tag{4.82b–d}$$

The particular integrals for the incremental displacement can then be determined by performing three differentiations of $\mathrm{d}h_{ij}$ and by substituting these into eqn (4.73):

$$\begin{aligned}\mathrm{d}u_i^{\mathrm{p}}(x) &= \sum_{n=1}^{N}\left[\frac{2(1+\nu)}{E}C_{,jll}(x, \xi_n)\delta_{ik} - \frac{(1+\nu)}{E(1-\nu)}C_{,jki}(x, \xi_n)\right]\mathrm{d}\psi_{kj}(\xi_n)\\ &= \sum_{n=1}^{N} D_{ijk}\mathrm{d}\psi_{kj}(\xi_n),\end{aligned} \tag{4.83}$$

where

$$D_{ijk}(x, \xi_n) = A_0^4\left[(c_1 + d_1\rho)(y_i\delta_{kj} + y_j\delta_{ik}) + (c_2 + d_2\rho)y_k\delta_{ij} + \frac{d_1}{\rho}y_i y_k y_j\right], \tag{4.84a}$$

$$c_1 = \frac{-8(1+\nu)}{E(1-\nu)}, \qquad c_2 = c_1 + \frac{16(d+2)(1+\nu)}{E}, \tag{4.84b,c}$$

$$d_1 = \frac{15b_{\mathrm{n}}(1+\nu)}{E(1-\nu)}, \qquad d_2 = d_1 - \frac{30b_{\mathrm{n}}(d+3)(1+\nu)}{E}. \tag{4.84d,e}$$

Substituting the above equation into the strain–displacement equation (4.74), leads to the particular integrals for the incremental strain:

$$\mathrm{d}\varepsilon_{ij}^{\mathrm{p}}(x) = \frac{1}{2}\sum_{n=1}^{N}\left[D_{ikl,j}(x, \xi_n) + D_{jkl,i}(x, \xi_n)\right]\mathrm{d}\psi_{kl}(\xi_n). \tag{4.85}$$

Subsequently, an expression for the particular integrals for the incremental stress (eqn (4.77)) are derived using the above equation and eqn (4.81):

$$\begin{aligned}\mathrm{d}\sigma_{ij}^{\mathrm{p}}(x) &= \sum_{n=1}^{N}\left[\frac{1}{2}D_{ijpq}^{\mathrm{e}}\left[D_{pkl,q}(x, \xi_n) + D_{qkl,p}(x, \xi_n)\right] - K(x, \xi_n)\delta_{ik}\delta_{jl}\right]\mathrm{d}\psi_{lk}(\xi_n)\\ &= \sum_{n=1}^{N} S_{ijlk}\mathrm{d}\psi_{lk}(\xi_n),\end{aligned} \tag{4.86}$$

in which

$$S_{ijlk}(x,\xi_n) = A_0^4\Big[(e_2+f_2\rho)\delta_{jk}\delta_{il}+(e_3+f_3\rho)\delta_{ij}\delta_{lk}+(e_4+f_4\rho)\delta_{ik}\delta_{jl} + \frac{f_1}{\rho}(y_jy_k\delta_{il}+y_iy_j\delta_{lk}+y_iy_k\delta_{jl})+\frac{f_2}{\rho}(y_iy_l\delta_{jk}+y_jy_l\delta_{ik}) + \frac{f_3}{\rho}y_ly_k\delta_{ij}+\frac{f_5}{\rho^3}y_iy_jy_ly_k\Big], \tag{4.87a}$$

where

$$e_1 = \frac{E}{(1+\nu)}c_1, \qquad e_2 = \frac{E}{2(1+\nu)}(c_1+c_2), \tag{4.87b,c}$$

$$e_3 = e_1 + \frac{E\nu}{(1+\nu)(1-2\nu)}\left[c_1(d+1)+c_2\right], \qquad e_4 = e_2 - a, \tag{4.87d,e}$$

$$f_1 = -f_5 = \frac{E}{(1+\nu)}d_1, \qquad f_2 = \frac{E}{2(1+\nu)}(d_1+d_2), \tag{4.87f,g}$$

$$f_3 = f_1 + \frac{E\nu}{(1+\nu)(1-2\nu)}\left[d_1(d+1)+d_2\right], \qquad f_4 = f_2 + b. \tag{4.87h,i}$$

Finally, the particular integrals for incremental traction can be found by substituting the above equation into eqn (4.78):

$$\mathrm{d}t_i^{\mathrm{p}}(x) = \mathrm{d}\sigma_{ij}^{\mathrm{p}}(x)n_j = \sum_{n=1}^{N}\left[S_{ijlk}(x,\xi_n)n_j\right]\mathrm{d}\psi_{lk}(\xi_n) = \sum_{n=1}^{N}\mathrm{T}_{ikl}(x,\xi_n)\mathrm{d}\psi_{lk}(\xi_n). \tag{4.88}$$

All the above equations are used in both the 3D analysis and for the plane strain case ($i, j = 1, 2$). When the plane stress condition is assumed ($i, j = 1, 2$), the same formulation is maintained simply by replacing E by $E(1+2\nu)/(1+\nu)^2$ and $\nu/(1+\nu)$.

Notice that the particular integrals will be applied in each region where plastic deformation is anticipated, independently of the other regions in a multi-region (zone) problem.

For numerical computational purposes, one needs to represent the particular integrals (solutions) for incremental displacement, traction, stress and also incremental initial stress evaluated from eqns (4.83), (4.88), (4.86) and (4.81), written in compact matrix forms instead of indicial notations. The first two particular integrals (solutions), which are the components of the incremental displacement and traction at all surface (boundary) nodes in the boundary element formulation, may be written:

$$\underset{(d\cdot M)\times(1)}{\{\mathrm{d}u^{\mathrm{p}}\}} = \underset{(d\cdot M)\times(p\cdot N)}{[D]} \cdot \underset{(p\cdot N)\times(1)}{\{\mathrm{d}\psi\}} \tag{4.89}$$

and

$$\underset{(d\cdot M)\times(1)}{\{\mathrm{d}t^{\mathrm{p}}\}} = \underset{(d\cdot M)\times(p\cdot N)}{[T]} \cdot \underset{(p\cdot N)\times(1)}{\{\mathrm{d}\psi\}}, \tag{4.90}$$

where the numbers in the parentheses underneath these equations and hereafter indicate the number of rows and columns present in the items they are closest to; d is the number of dimensions of the model, i.e. $d = 2$ in 2D and 3 in 3D, M the number of surface (boundary) nodes, p the number of independent stress components, i.e. $p = 3$ in 2D and 6 in 3D, and, N the sum of the number of surface (boundary) nodes and interior sample points. The entries in the above vectors in 3D can be explicitly expressed:

$$\{\{du^{\mathrm{p}}\}^{\mathrm{T}} = \{\{du_1(\xi_1)du_2(\xi_1)du_3(\xi_1)\} \quad \{du_1(\xi_2)du_2(\xi_2)du_3(\xi_2)\} \cdots \{du_1(\xi_M)du_2(\xi_M)du_3(\xi_M)\}\} \tag{4.91}$$

and

$$\begin{aligned}\{d\psi\}^{\mathrm{T}} =& \{\{d\psi_{11}(\xi_1)d\psi_{11}(\xi_2)\cdots d\psi_{11}(\xi_N)\} \quad \{d\psi_{22}(\xi_1)d\psi_{22}(\xi_2)\cdots d\psi_{22}(\xi_N)\}\\ &\{d\psi_{33}(\xi_1)d\psi_{33}(\xi_2)\cdots d\psi_{33}(\xi_N)\} \quad \{d\psi_{12}(\xi_1)d\psi_{12}(\xi_2)\cdots d\psi_{12}(\xi_N)\}\\ &\{d\psi_{23}(\xi_1)d\psi_{23}(\xi_2)\cdots d\psi_{23}(\xi_N)\} \quad \{d\psi_{31}(\xi_1)d\psi_{31}(\xi_2)\cdots d\psi_{31}(\xi_N)\}\}.\end{aligned} \tag{4.92}$$

It should be noted that, firstly, the column vector of the components of the particular integrals for incremental traction (eqn (4.90)) has the same structure as the incremental displacement shown in eqn (4.91). Secondly, eqns (4.90, 4.91), which are presented in 3D, can be given in 2D by considering two components of $\{du^{\mathrm{p}}\}$ for M nodes and three combinations of the subscripts of $\{d\xi\}^{\mathrm{T}}$ such as $\{\{d\xi_{11}\}\{d\xi_{22}\}\{d\xi_{12}\}\}$ for N nodes and sample points. Finally, in eqns (4.83) and (4.88), which are used for the determination of the coefficients of $[D]$ and $[T]$, x is replaced by ξ_n, where $n = 1, 2, \ldots, M$.

The matrix form of the particular integrals (solutions) for the components of the incremental stress (eqn (4.86)) can be expressed:

$$\underset{(p\cdot N)\times(1)}{\{d\sigma^{\mathrm{p}}\}} = \underset{(p\cdot N)\times(p\cdot N)}{[S]} \cdot \underset{(p\cdot N)\times(1),}{\{d\psi\}} \tag{4.93}$$

in which:

(a) In 2D:

$$\{d\sigma^{\mathrm{p}}\}^{\mathrm{T}} = \{\{d\sigma_{11}\} \quad \{d\sigma_{22}\} \quad \{d\sigma_{12}\}\}. \tag{4.94a}$$

(b) In 3D:

$$\{d\sigma^{\mathrm{p}}\}^{\mathrm{T}} = \{\{d\sigma_{11}\} \quad \{d\sigma_{22}\} \quad \{d\sigma_{33}\} \quad \{d\sigma_{12}\} \quad \{d\sigma_{23}\} \quad \{d\sigma_{31}\}\}, \tag{4.94b}$$

which $\{d\sigma^{\mathrm{p}}\}^{\mathrm{T}}$ have the same structure as $\{d\psi\}^{\mathrm{T}}$ shown in eqn (4.92).

Similarly, the incremental initial stress (eqn (4.81)) may be represented in matrix form:

$$\underset{(p\cdot N)\times(1)}{\{d\sigma^{\mathrm{o}}\}} = \underset{(p\cdot N)\times(p\cdot N)}{[K]} \cdot \underset{(p\cdot N)\times(1).}{\{d\psi\}} \tag{4.95}$$

It should be noted that because the indices i and j in eqn (4.84) are free and also $\{d\psi\}^{\mathrm{T}}$, shown in eqn (4.92), has the particular order of the components, matrix $[K]$ takes on the block diagonal structure shown below.

(a) In 2D:

$$\begin{Bmatrix}\{d\sigma^{\mathrm{o}}_{11}\}\\ \{d\sigma^{\mathrm{o}}_{22}\}\\ \{d\sigma^{\mathrm{o}}_{12}\}\end{Bmatrix} = \begin{bmatrix}[K_1] & [0] & [0]\\ [0] & [K_1] & [0]\\ [0] & [0] & [K_1]\end{bmatrix} \begin{Bmatrix}\{d\psi_{11}\}\\ \{d\psi_{22}\}\\ \{d\psi_{12}\}\end{Bmatrix}. \tag{4.96a}$$

(b) In 3D:

$$\begin{Bmatrix} \{\mathrm{d}\sigma_{11}^{\mathrm{o}}\} \\ \{\mathrm{d}\sigma_{22}^{\mathrm{o}}\} \\ \{\mathrm{d}\sigma_{33}^{\mathrm{o}}\} \\ \{\mathrm{d}\sigma_{12}^{\mathrm{o}}\} \\ \{\mathrm{d}\sigma_{23}^{\mathrm{o}}\} \\ \{\mathrm{d}\sigma_{31}^{\mathrm{o}}\} \end{Bmatrix} = \begin{bmatrix} [K_1] & [0] & [0] & [0] & [0] & [0] \\ [0] & [K_1] & [0] & [0] & [0] & [0] \\ [0] & [0] & [K_1] & [0] & [0] & [0] \\ [0] & [0] & [0] & [K_1] & [0] & [0] \\ [0] & [0] & [0] & [0] & [K_1] & [0] \\ [0] & [0] & [0] & [0] & [0] & [K_1] \end{bmatrix} \begin{Bmatrix} \{\mathrm{d}\psi_{11}\} \\ \{\mathrm{d}\psi_{22}\} \\ \{\mathrm{d}\psi_{33}\} \\ \{\mathrm{d}\psi_{12}\} \\ \{\mathrm{d}\psi_{23}\} \\ \{\mathrm{d}\psi_{31}\} \end{Bmatrix}. \tag{4.96b}$$

The different variables of the particular integrals (solutions), obtained by the above series of equations in indicial notations as well as in matrix forms, have been expressed in terms of the unknown function $\{\mathrm{d}\psi\}$. This function can be represented by making use of eqn (4.95):

$$\{\mathrm{d}\psi\} = [K]^{-1}\{\mathrm{d}\sigma^{\mathrm{o}}\}, \tag{4.97}$$

where an $[L][U]$ factorization of $[K]$ (actually $[K_1]$) is applied in practice, rather than calculating the inverse matrix.

Substitution of the above equation into eqns (4.89, 4.90) and (4.93) yields:

$$\{\mathrm{d}u^{\mathrm{p}}\} = [D][K]^{-1}\{\mathrm{d}\sigma^{\mathrm{o}}\}, \tag{4.98}$$

$$\{\mathrm{d}t^{\mathrm{p}}\} = [\mathrm{T}][K]^{-1}\{\mathrm{d}\sigma^{\mathrm{o}}\} \tag{4.99}$$

and

$$\{\mathrm{d}\sigma^{\mathrm{p}}\} = [S][K]^{-1}\{\mathrm{d}\sigma^{\mathrm{o}}\}. \tag{4.100}$$

Furthermore, the last equation, combined with eqn (4.77), yields an expression for the incremental initial stress in terms of the particular integrals for the incremental strain:

$$\begin{aligned} \{\mathrm{d}\sigma^{\mathrm{o}}\} &= \Big[[I] + [S][K]^{-1}\Big]^{-1} \cdot [D^{\mathrm{e}}]\{\mathrm{d}\varepsilon^{\mathrm{p}}\} \\ &\equiv [D_{\mathrm{p}}]\{\mathrm{d}\varepsilon^{\mathrm{p}}\}, \end{aligned} \tag{4.101}$$

where $[I]$ is the identity matrix and

$$[D_{\mathrm{p}}] = \Big[[I] + [S][K]^{-1}\Big]^{-1} \cdot [D^{\mathrm{e}}]. \tag{4.102}$$

Substituting these relations into eqn (4.95) leads to

$$\begin{aligned} \{\mathrm{d}\psi\} &= [K]^{-1} \cdot \Big[[I] + [S][K]^{-1}\Big]^{-1} \cdot [D^{\mathrm{e}}]\{\mathrm{d}\varepsilon^{\mathrm{p}}\} \\ &\equiv [K]^{-1}[D_{\mathrm{p}}]\{\mathrm{d}\varepsilon^{\mathrm{p}}\}. \end{aligned} \tag{4.103}$$

Finally, the boundary element formulation for elastoplastic problems based on the concept of particular integrals (solutions) is obtained by substituting eqns (4.98–4.101) into eqns (4.64, 4.65):

$$\begin{aligned} [H]\{\mathrm{d}u\} &= [G]\{\mathrm{d}t\} - [[G][\mathrm{T}] - [H][D]]\,[K]^{-1}\{\mathrm{d}\sigma^{\mathrm{o}}\} \\ &= [G]\{\mathrm{d}t\} + [B]\{\mathrm{d}\sigma^{\mathrm{o}}\}, \end{aligned} \tag{4.104}$$

$$\begin{aligned} \{\mathrm{d}\sigma\} &= [G^{\sigma}]\{\mathrm{d}t\} - [H^{\sigma}]\{\mathrm{d}u\} - [[G^{\sigma}][\mathrm{T}] - [H^{\sigma}][D] - [S]][K]^{-1}\{\mathrm{d}\sigma^{\mathrm{o}}\} \\ &= [G^{\sigma}]\{\mathrm{d}t\} - [H^{\sigma}]\{\mathrm{d}u\} + [B^{\sigma}]\{\mathrm{d}\sigma^{\mathrm{o}}\}, \end{aligned} \tag{4.105}$$

where

$$[B] = -[[G][\mathrm{T}] - [H][D]][K]^{-1}, \tag{4.106}$$

$$[B^{\sigma}] = -[[G^{\sigma}][\mathrm{T}] - [H^{\sigma}][D] - [S]][K]^{-1}. \tag{4.107}$$

It can be shown that the above system of equations is structurally identical in form to the one derived by using the approach that requires volume integration, and therefore, its solution may be obtained by employing any elastoplastic solution procedure (Henry and Banerjee, 1988).

In the present study, in order to take advantage of the linear complementarity problem (LCP), the above boundary element formulation based on the particular integrals concept for incremental initial strain, rather than the incremental initial stress, will be employed. Substituting eqn (4.101) into the above equations we obtain:

$$[H]\{\mathrm{d}u\} = [G]\{\mathrm{d}t\} + [B][D_{\mathrm{p}}]\{\mathrm{d}\varepsilon^{\mathrm{p}}\}, \tag{4.108}$$

$$\{\mathrm{d}\sigma\} = [G^{\sigma}]\{\mathrm{d}t\} - [H^{\sigma}]\{\mathrm{d}u\} + [B^{\sigma}][D_{\mathrm{p}}]\{\mathrm{d}\varepsilon^{\mathrm{p}}\}. \tag{4.109}$$

We will now attempt to employ the boundary element formulation for elastoplastic contact problems, which provides the necessary and sufficient relations between the contact variables and the stress field in incremental form for contacting bodies. These will then be the last supplementary equations needed to solve the proposed elastoplastic frictional contact problem presented in eqns (4.4, 4.5), (4.20) and (4.24, 4.25). For each of the contacting bodies $\alpha(\alpha = 1, 2)$ described in Section 2.2 of Chapter 2, the boundary element formulation (eqns (4.108, 4.109)) can be rewritten:

$$[H]^{\alpha}\{\mathrm{d}u\}^{\alpha} = [G]^{\alpha}\{\mathrm{d}t\}^{\alpha} + [B]^{\alpha}[D_{\mathrm{p}}]^{\alpha}\{\mathrm{d}\varepsilon^{\mathrm{p}}\}^{\alpha}, \tag{4.110}$$

$$\{\mathrm{d}\sigma\}^{\alpha} = [G^{\sigma}]^{\alpha}\{\mathrm{d}t\}^{\alpha} - [H^{\sigma}]^{\alpha}\{\mathrm{d}u\}^{\alpha} + [B^{\sigma}]^{\alpha}[D_{\mathrm{p}}]^{\alpha}\{\mathrm{d}\varepsilon^{\mathrm{p}}\}^{\alpha}. \tag{4.111}$$

By applying the boundary conditions with respect to points inside and outside the potential contact surface S_{c} and rearranging the corresponding columns and rows in the matrices $[H]$ and $[G]$ as well as $[H^{\sigma}]$ and $[G^{\sigma}]$, one finds, at the ith incremental step, the following equations:

$$[H \quad H_{\mathrm{c}}]^{\alpha}\left(\begin{Bmatrix}\mathrm{d}u\\ \mathrm{d}u_{\mathrm{c}}\end{Bmatrix}_{\mathrm{g}}^{\alpha}\right)_i = [G \quad G_{\mathrm{c}}]^{\alpha}\left(\begin{Bmatrix}\Gamma\mathrm{d}\bar{t}\\ \mathrm{d}p_{\mathrm{c}}\end{Bmatrix}_{\mathrm{g}}^{\alpha}\right)_i + [B]^{\alpha}[D_{\mathrm{p}}]^{\alpha}\left\{\mathrm{d}\varepsilon^{\mathrm{p}}\right\}_i^{\alpha}, \tag{4.112}$$

$$\{\mathrm{d}\sigma\}_i^{\alpha} = \left[G^{\sigma} \quad G_{\mathrm{c}}^{\sigma}\right]^{\alpha}\left(\begin{Bmatrix}\Gamma\mathrm{d}\bar{t}\\ \mathrm{d}p_{\mathrm{c}}\end{Bmatrix}_{\mathrm{g}}^{\alpha}\right)_i - \left[H^{\sigma} \quad H_{\mathrm{c}}^{\sigma}\right]^{\alpha}\left(\begin{Bmatrix}\mathrm{d}u\\ \mathrm{d}u_{\mathrm{c}}\end{Bmatrix}_{\mathrm{g}}^{\alpha}\right)_i + [B^{\sigma}]^{\alpha}[D_{\mathrm{p}}]^{\alpha}\left\{\mathrm{d}\varepsilon^{\mathrm{p}}\right\}_i^{\alpha}, \tag{4.113}$$

where $\{\mathrm{d}u\}_{\mathrm{g}}^{\alpha}$ and $\{\Gamma\mathrm{d}\bar{t}\}_{\mathrm{g}}^{\alpha}$ represent unknown and known incremental boundary data outside the potential contact surface S_{c} respectively, while $\{\mathrm{d}u_{\mathrm{c}}\}_{\mathrm{g}}^{\alpha}$ and $\{\mathrm{d}p_{\mathrm{c}}\}_{\mathrm{g}}^{\alpha}$ represent the unknown incremental contact displacements and tractions. The subscript g indicates that the equation is expressed in the global coordinate system.

Starting with eqn (4.112) and multiplying the inverse of matrix $[H \quad H_{\mathrm{c}}]^{\alpha}$ with matrices $[G \quad G_{\mathrm{c}}]^{\alpha}$ and $[[B]^{\alpha}[D_{\mathrm{p}}]^{\alpha}]$, these global systems of equations become:

$$\begin{aligned}\left(\begin{Bmatrix}\mathrm{d}u\\ \mathrm{d}u_{\mathrm{c}}\end{Bmatrix}_{\mathrm{g}}^{\alpha}\right)_i &= \left[[H \quad H_{\mathrm{c}}]^{-1}[G \quad G_{\mathrm{c}}]\right]^{\alpha}\left(\begin{Bmatrix}\Gamma\mathrm{d}\bar{t}\\ \mathrm{d}p_{\mathrm{c}}\end{Bmatrix}_{\mathrm{g}}^{\alpha}\right)_i + \left[[H \quad H_{\mathrm{c}}]^{-1}[B][D_{\mathrm{p}}]\right]^{\alpha}\left\{\mathrm{d}\varepsilon^{\mathrm{p}}\right\}_i^{\alpha} \\ &= \begin{bmatrix}f_{\mathrm{oo}} & f_{\mathrm{oc}}\\ f_{\mathrm{co}} & f_{\mathrm{cc}}\end{bmatrix}^{\alpha}\left(\begin{Bmatrix}\Gamma\mathrm{d}\bar{t}\\ \mathrm{d}p_{\mathrm{c}}\end{Bmatrix}_{\mathrm{g}}^{\alpha}\right)_i + \begin{bmatrix}C_{\mathrm{o}}^{u}\\ C_{\mathrm{c}}^{u}\end{bmatrix}^{\alpha}\{\mathrm{d}\varepsilon^{\mathrm{p}}\}_i^{\alpha},\end{aligned} \tag{4.114}$$

where $\begin{bmatrix} f_{\text{oo}} & f_{\text{oc}} \\ f_{\text{co}} & f_{\text{cc}} \end{bmatrix}^{\alpha}$ is the flexibility matrix and the subscripts 'o' and 'c' indicate outside and inside the potential contact zone respectively.

This equation immediately provides the required relation between the contact displacements and the contact tractions, external loading $\{\Gamma \mathrm{d}\bar{t}\}_{\mathrm{g}}^{\alpha}$ and plastic strain increment vector, as well as an equation to find the displacement components outside the contact surface. In explicit form, these can be written:

$$\left(\{\mathrm{d}u_{\mathrm{c}}\}_{\mathrm{g}}^{\alpha}\right)_i = [f_{\mathrm{co}}]^{\alpha}\left(\left\{\Gamma \mathrm{d}\bar{t}\right\}_{\mathrm{g}}^{\alpha}\right)_i + [f_{\mathrm{cc}}]^{\alpha}\left(\{\mathrm{d}p_{\mathrm{c}}\}_{\mathrm{g}}^{\alpha}\right)_i + \left[C_{\mathrm{c}}^{u}\right]^{\alpha}\left\{\mathrm{d}\varepsilon^{\mathrm{p}}\right\}_i^{\alpha} \tag{4.115}$$

and

$$\left(\{\mathrm{d}u\}_{\mathrm{g}}^{\alpha}\right)_i = [f_{\mathrm{oo}}]^{\alpha}\left(\left\{\Gamma \mathrm{d}\bar{t}\right\}_{\mathrm{g}}^{\alpha}\right)_i + [f_{\mathrm{oc}}]^{\alpha}\left(\{\mathrm{d}p_{\mathrm{c}}\}_{\mathrm{g}}^{\alpha}\right)_i + \left[C_{\mathrm{o}}^{u}\right]^{\alpha}\left\{\mathrm{d}\varepsilon^{\mathrm{p}}\right\}_i^{\alpha}. \tag{4.116}$$

Notice that once the incremental contact tractions $\{\mathrm{d}p_{\mathrm{c}}\}_{\mathrm{g}}^{\alpha}$ have been found, $\{\mathrm{d}u\}_{\mathrm{g}}^{\alpha}$ can be obtained immediately from eqn (4.116).

The contact variables described in Section 2.2 of Chapter 2, were presented in a local coordinate system. It is thus necessary to transform the variables in eqn (4.115) into the local coordinate system by means of the transformation matrix [Tr] defined in eqns (2.76b,c). One therefore has:

$$\begin{aligned} \left(\{\mathrm{d}u_{\mathrm{c}}\}_{\mathrm{g}}^{\alpha}\right)_i &= [\mathrm{Tr}]\,\{\mathrm{d}u_{\mathrm{c}}\}_i^{\alpha}, \\ \left(\{\mathrm{d}p_{\mathrm{c}}\}_{\mathrm{g}}^{\alpha}\right)_i &= [\mathrm{Tr}]\,\{\mathrm{d}p_{\mathrm{c}}\}_i^{\alpha}. \end{aligned} \tag{4.117}$$

Note that the local incremental contact variables $\{\mathrm{d}u_{\mathrm{c}}\}_i^{\alpha}$ and $\{\mathrm{d}p_{\mathrm{c}}\}_i^{\alpha}$ are shown without subscript ℓ for simplicity.

Therefore, eqn (4.115) for the contact point pairs becomes:

$$\{\mathrm{d}u_{\mathrm{c}}\}_i^{\alpha} = \{F\}_i^{\alpha} + [A]^{\alpha}\{\mathrm{d}p_{\mathrm{c}}\}_i^{\alpha} + \left[B^{u}\right]^{\alpha}\left\{\mathrm{d}\varepsilon^{\mathrm{p}}\right\}_i^{\alpha}, \tag{4.118}$$

where

$$\begin{aligned} \{F\}_i^{\alpha} &= [\mathrm{Tr}]^{-1}\{f_{\mathrm{co}}\}^{\alpha}\left(\left\{\Gamma \mathrm{d}\bar{t}\right\}_{\mathrm{g}}^{\alpha}\right)_i, \\ [A]^{\alpha} &= [\mathrm{Tr}]^{-1}\{f_{\mathrm{cc}}\}^{\alpha}[\mathrm{Tr}], \\ [B^{u}]^{\alpha} &= [\mathrm{Tr}]^{-1}\left[C_{\mathrm{c}}^{u}\right]^{\alpha}. \end{aligned} \tag{4.119}$$

In accordance with the normal and tangential components of the incremental contact variables (eqn (4.118)) can be expressed explicitly:

(a) In 2D:

$$\begin{Bmatrix} \mathrm{d}u_{\mathrm{cn}} \\ \mathrm{d}u_{\mathrm{ct}} \end{Bmatrix}_i^{\alpha} = \begin{Bmatrix} F_{\mathrm{n}} \\ F_{\mathrm{t}} \end{Bmatrix}_i^{\alpha} + [A_{\mathrm{n}} \quad A_{\mathrm{t}}]^{\alpha}\begin{Bmatrix} \mathrm{d}p_{\mathrm{cn}} \\ \mathrm{d}p_{\mathrm{ct}} \end{Bmatrix}_i^{\alpha} + \begin{bmatrix} B_{\mathrm{n}}^{u} \\ B_{\mathrm{t}}^{u} \end{bmatrix}^{\alpha}\left\{\mathrm{d}\varepsilon^{\mathrm{p}}\right\}_i^{\alpha}. \tag{4.120}$$

(b) In 3D:

$$\begin{Bmatrix} \mathrm{d}u_{\mathrm{cn}} \\ \mathrm{d}u_{\mathrm{ct_1}} \\ \mathrm{d}u_{\mathrm{ct_2}} \end{Bmatrix}_i^{\alpha} = \begin{Bmatrix} F_{\mathrm{n}} \\ F_{\mathrm{t1}} \\ F_{\mathrm{t2}} \end{Bmatrix}_i^{\alpha} + [A_{\mathrm{n}} \quad A_{\mathrm{t1}} \quad A_{\mathrm{t2}}]^{\alpha}\begin{Bmatrix} \mathrm{d}p_{\mathrm{cn}} \\ \mathrm{d}p_{\mathrm{ct_1}} \\ \mathrm{d}p_{\mathrm{ct_2}} \end{Bmatrix}_i^{\alpha} + \begin{bmatrix} B_{\mathrm{n}}^{u} \\ B_{\mathrm{t1}}^{u} \\ B_{\mathrm{t2}}^{u} \end{bmatrix}^{\alpha}\left\{\mathrm{d}\varepsilon^{\mathrm{p}}\right\}_i^{\alpha}. \tag{4.121}$$

Substituting eqns (4.118–4.121) into eqn (2.3) of Chapter 2, Section 2.2, and by considering its incremental forms, one finds the following required relation between the incremental contact displacements and contact tractions, in addition to the elastoplastic term, derived from the boundary element formulations:

$$\{\mathrm{d}u_{\mathrm{c}}\}_i = \{F_{\mathrm{c}}\}_i + [A_{\mathrm{c}}]\{\mathrm{d}p_{\mathrm{c}}\}_i + [B^u]\{\mathrm{d}\varepsilon^{\mathrm{p}}\}_i, \tag{4.122}$$

where

$$\{\mathrm{d}\varepsilon^{\mathrm{p}}\}_i = \begin{Bmatrix} \{\mathrm{d}\varepsilon^{\mathrm{p}}\}_i^1 \\ \{\mathrm{d}\varepsilon^{\mathrm{p}}\}_i^2 \end{Bmatrix}. \tag{4.123}$$

(a) In 2D:

$$\begin{aligned}
&\{\mathrm{d}u_{\mathrm{c}}\}_i = \begin{Bmatrix} \mathrm{d}u_{\mathrm{cn}} \\ \mathrm{d}u_{\mathrm{ct}} \end{Bmatrix}_i \equiv \begin{Bmatrix} \mathrm{d}u_{\mathrm{cn}}^2 - \mathrm{d}u_{\mathrm{cn}}^1 + u_0 \\ \mathrm{d}u_{\mathrm{ct}}^2 - \mathrm{d}u_{\mathrm{ct}}^1 \end{Bmatrix}_i, \\
&\{F_{\mathrm{c}}\}_i = \begin{Bmatrix} F_{\mathrm{cn}} \\ F_{\mathrm{ct}} \end{Bmatrix}_i = \begin{Bmatrix} F_{\mathrm{n}}^2 - F_{\mathrm{n}}^1 + u_0 \\ F_{\mathrm{t}}^2 - F_{\mathrm{t}}^1 \end{Bmatrix}_i, \\
&[A_{\mathrm{c}}] = [A_{\mathrm{cn}} \quad A_{\mathrm{ct}}] = [-(A_{\mathrm{n}}^1 + A_{\mathrm{n}}^2)(A_{\mathrm{t}}^1 + A_{\mathrm{t}}^2)], \\
&\{\mathrm{d}p_{\mathrm{c}}\}_i = \begin{Bmatrix} -\mathrm{d}p_{\mathrm{cn}} \\ \mathrm{d}p_{\mathrm{ct}} \end{Bmatrix}_i, \\
&[B^u] = \begin{bmatrix} -[B_{\mathrm{n}}^u]^1 & [B_{\mathrm{n}}^u]^2 \\ -[B_{\mathrm{t}}^u]^1 & [B_{\mathrm{t}}^u]^2 \end{bmatrix}.
\end{aligned} \tag{4.124}$$

(b) In 3D:

$$\begin{aligned}
&\{\mathrm{d}u_{\mathrm{c}}\}_i = \begin{Bmatrix} \mathrm{d}u_{\mathrm{cn}} \\ \mathrm{d}u_{\mathrm{ct}_1} \\ \mathrm{d}u_{\mathrm{ct}_2} \end{Bmatrix}_i \equiv \begin{Bmatrix} \mathrm{d}u_{\mathrm{cn}}^2 - \mathrm{d}u_{\mathrm{cn}}^1 + u_0 \\ \mathrm{d}u_{\mathrm{ct}_1}^2 - \mathrm{d}u_{\mathrm{ct}_1}^1 \\ \mathrm{d}u_{\mathrm{ct}_2}^2 - \mathrm{d}u_{\mathrm{ct}_2}^1 \end{Bmatrix}_i, \\
&\{F_{\mathrm{c}}\}_i = \begin{Bmatrix} F_{\mathrm{cn}} \\ F_{\mathrm{ct}_1} \\ F_{\mathrm{ct}_2} \end{Bmatrix} = \begin{Bmatrix} F_{\mathrm{n}}^2 - F_{\mathrm{n}}^1 + u_0 \\ F_{\mathrm{t1}}^2 - F_{\mathrm{t1}}^1 \\ F_{\mathrm{t2}}^2 - F_{\mathrm{t2}}^1 \end{Bmatrix}_i, \\
&[A_{\mathrm{c}}] = [A_{\mathrm{cn}} \quad A_{\mathrm{ct}}] = \left[-(A_{\mathrm{n}}^1 + A_{\mathrm{n}}^2) \quad (A_{\mathrm{t1}}^1 + A_{\mathrm{t1}}^2) \quad (A_{\mathrm{t2}}^1 + A_{\mathrm{t2}}^2)\right], \\
&\{\mathrm{d}p_{\mathrm{c}}\}_i = \begin{Bmatrix} -\mathrm{d}p_{\mathrm{cn}} \\ \mathrm{d}p_{\mathrm{ct}_1} \\ \mathrm{d}p_{\mathrm{ct}_2} \end{Bmatrix}_i, \\
&[B^u] = \begin{bmatrix} -\left[B_{\mathrm{n}}^u\right]^1 & \left[B_{\mathrm{n}}^u\right]^2 \\ -\left[B_{\mathrm{t1}}^u\right]^1 & \left[B_{\mathrm{t1}}^u\right]^2 \\ -\left[B_{\mathrm{t2}}^u\right]^1 & \left[B_{\mathrm{t2}}^u\right]^2 \end{bmatrix}.
\end{aligned} \tag{4.125}$$

Therefore, the application of the boundary element formulation for elastoplastic contact problems yields eqns (4.122–4.125), which provide the necessary supplementary equations for calculating the unknown contact variables introduced in eqns (4.4, 4.5). We will now describe the procedure to find the missing equations.

In order to find the last supplementary equation to solve the proposed elastoplastic frictional contact problem defined by eqns (4.4, 4.5), (4.20) and (4.24, 4.25), one can start by substituting eqn (4.116) into eqn (4.113), which yields after some manipulations:

$$\begin{aligned}\{\mathrm{d}\sigma\}_i^\alpha = {} & \left[G^\sigma - H^\sigma f_{\mathrm{oo}}\right]^\alpha \left(\{\Gamma \mathrm{d}\bar{t}\}_{\mathrm{g}}^\alpha\right)_i + \left[G_{\mathrm{c}}^\sigma - H^\sigma f_{\mathrm{oc}}\right]^\alpha \left(\{\mathrm{d}p_{\mathrm{c}}\}_{\mathrm{g}}^\alpha\right)_i \\ & - \left[H_{\mathrm{c}}^\sigma\right]^\alpha \left(\{\mathrm{d}u_{\mathrm{c}}\}_{\mathrm{g}}^\alpha\right)_i + \left[-H^\sigma C_{\mathrm{o}}^u + B^\sigma D_{\mathrm{p}}\right]^\alpha \left\{\mathrm{d}\varepsilon^{\mathrm{p}}\right\}_i^\alpha .\end{aligned} \tag{4.126}$$

Substituting the contact variables given in eqn (4.117) into the above equation, one gets, after simplification:

$$\{\mathrm{d}\sigma\}_i^\alpha = \{R_0\}_i^\alpha + [R_1]^\alpha \{\mathrm{d}p_{\mathrm{c}}\}_i^\alpha + [R_2]^\alpha \{\mathrm{d}u_{\mathrm{c}}\}_i^\alpha + [R_3]^\alpha \left\{\mathrm{d}\varepsilon^{\mathrm{p}}\right\}_i^\alpha , \tag{4.127}$$

where

$$\{R_0\}_i^\alpha = \left[G^\sigma - H^\sigma f_{\mathrm{oo}}\right]^\alpha \left(\left\{\Gamma \mathrm{d}\bar{t}\right\}_{\mathrm{g}}^\alpha\right)_i , \tag{4.128a}$$

$$[R_1]^\alpha = \left[G_{\mathrm{c}}^\sigma - H^\sigma f_{\mathrm{oc}}\right]^\alpha [\mathrm{T}r], \tag{4.128b}$$

$$[R_2]^\alpha = -\left[H_{\mathrm{c}}^\sigma\right]^\alpha [\mathrm{T}r], \tag{4.128c}$$

$$[R_3]^\alpha = \left[-H^\sigma C_{\mathrm{o}}^u + B^\sigma D_{\mathrm{p}}\right]^\alpha . \tag{4.128d}$$

Replacing the incremental contact displacements in the above equation by those presented in eqn (4.118) yields:

$$\{\mathrm{d}\sigma\}_i^\alpha = \{R\}_i^\alpha + [R_{\mathrm{c}}]^\alpha \{\mathrm{d}p_{\mathrm{c}}\}_i^\alpha + [R^\sigma]^\alpha \{\mathrm{d}\varepsilon^{\mathrm{p}}\}_i^\alpha , \tag{4.129}$$

in which

$$\{R\}_i^\alpha = \{R_0\}_i^\alpha + [R_2]^\alpha \{F\}_i^\alpha , \tag{4.130a}$$

$$[R_{\mathrm{c}}]^\alpha = [R_1]^\alpha + [R_2]^\alpha [A]^\alpha , \tag{4.130b}$$

$$[R^\sigma]^\alpha = [R_2]^\alpha [B^u]^\alpha + [R_3]^\alpha . \tag{4.130c}$$

Finally, by recalling eqn (2.6) and eqns (4.124, 4.125) for $\{\mathrm{d}p_{\mathrm{c}}\}_i$ and eqn (4.123) for $\{\mathrm{d}\varepsilon^{\mathrm{p}}\}_i$, the assembled form of the above two bodies contact equations corresponding to eqns (4.24, 4.25) is obtained:

$$\{\mathrm{d}\sigma\}_i = \{F^\sigma\}_i + [A_{\mathrm{c}}^\sigma]\{\mathrm{d}p_{\mathrm{c}}\}_i + [R^\sigma]\{\mathrm{d}\varepsilon^{\mathrm{p}}\}_i , \tag{4.131}$$

where

$$\{\mathrm{d}\sigma\}_i = \left\{ \begin{matrix} \{\mathrm{d}\sigma\}_i^1 \\ \{\mathrm{d}\sigma\}_i^2 \end{matrix} \right\} , \tag{4.132a}$$

$$\{F^\sigma\}_i = \left\{ \begin{matrix} \{R\}_i^1 \\ \{R\}_i^2 \end{matrix} \right\} , \tag{4.132b}$$

$$[A_{\mathrm{c}}^\sigma] = \begin{bmatrix} [R_{\mathrm{cn}}]^1 & -[R_{\mathrm{ct}}]^1 \\ -[R_{\mathrm{cn}}]^2 & [R_{\mathrm{ct}}]^2 \end{bmatrix} , \tag{4.132c}$$

$$[R^\sigma] = \begin{bmatrix} [R^\sigma]^1 & 0 \\ 0 & [R^\sigma]^2 \end{bmatrix} . \tag{4.132d}$$

At this point, we have obtained all the necessary and sufficient relations between the contact variables and the stress field in incremental form, as given by eqns (4.122) and (4.131), using the BEM. The procedure to find the unknown variables of the problem is presented in the next section.

4.6 Unified linear complementarity formulation of elastoplastic frictional contact problems

We now attempt to derive a unified mathematical programming approach for the elastoplastic frictional contact problems presented in the previous sections. According to the last section, the important and necessary supplementary eqns (4.122) and (4.131) are found to obtain the unknown variables of the problem defined in eqns (4.4, 4.5), (4.20) and (4.24, 4.25).

In eqn (4.122), we introduce eqn (4.3) for $\{\mathrm{d}p_\mathrm{c}\}_i$ and eqn (4.25) for $\{\mathrm{d}\varepsilon^\mathrm{p}\}_i$. After some manipulations, we obtain an equation for the incremental contact displacements $\{\mathrm{d}u_\mathrm{c}\}$ in terms of both the contact multipliers $\{\mathrm{d}\lambda_\mathrm{c}\}$ and the plastic multipliers $\{\mathrm{d}\lambda_\mathrm{p}\}$ with respect to the ith incremental step:

$$\{\mathrm{d}u_\mathrm{c}\}_i = \{E_\mathrm{c}\}_i + [B_\mathrm{c}]\{\mathrm{d}\lambda_\mathrm{c}\}_i + [B_\mathrm{p}]_{i-1}\left\{\mathrm{d}\lambda_\mathrm{p}\right\}_i, \tag{4.133}$$

where

$$\begin{aligned}
\{E_\mathrm{c}\}_i &= [[I]-[A_\mathrm{c}][D_\mathrm{c}]]^{-1}\cdot\{\{F_\mathrm{c}\}_i + [A_\mathrm{c}]\{[D_\mathrm{c}]\{u_\mathrm{c}\}_{i-1} - \{p_\mathrm{c}\}_{i-1}\}\},\\
[B_\mathrm{c}] &= -[[I]-[A_\mathrm{c}][D_\mathrm{c}]]^{-1}\cdot[[A_\mathrm{c}][D_\mathrm{c}][C_\mathrm{c}]],\\
[B_\mathrm{p}]_{i-1} &= [[I]-[A_\mathrm{c}][D_\mathrm{c}]]^{-1}\cdot[[B^u][C_\mathrm{p}]^\mathrm{T}_{i-1}],
\end{aligned}\tag{4.134}$$

in which $[I]$ is the identity matrix.

The first linear complementarity problem (LCP) for elastoplastic frictional contact problems may then be derived, by substitution of the above equations into eqn (4.4), yielding:

$$\begin{aligned}
-\{f_\mathrm{c}\}_i &= \{q_\mathrm{c}\}_i + [M_\mathrm{cc}]\{\mathrm{d}\lambda_\mathrm{c}\}_i + [M_\mathrm{cp}]_{i-1}\{\mathrm{d}\lambda_\mathrm{p}\}_i\\
&= \{q_\mathrm{c}\}_i + [[M_\mathrm{cc}][M_\mathrm{cp}]_{i-1}]\begin{Bmatrix}\{\mathrm{d}\lambda_\mathrm{c}\}_i\\ \{\mathrm{d}\lambda_\mathrm{p}\}_i\end{Bmatrix}\\
&= \{q_\mathrm{c}\}_i + [[M_\mathrm{cc}][M_\mathrm{cp}]_{i-1}]\{\mathrm{d}\lambda\}_i,
\end{aligned}\tag{4.135}$$

where

$$\begin{aligned}
\{q_\mathrm{c}\}_i &= [[C_\ell][D_\mathrm{c}]]\cdot\{\{E_\mathrm{c}\}_i + \{u_\mathrm{c}\}_{i-1}\},\\
[M_\mathrm{cc}] &= [[C_\ell][D_\mathrm{c}]]\cdot[[B_\mathrm{c}] - [C_\mathrm{c}]],\\
[M_\mathrm{cp}]_{i-1} &= [[C_\ell][D_\mathrm{c}]]\cdot[B_\mathrm{p}]_{i-1}.
\end{aligned}\tag{4.136}$$

The last linear complementarity problem (LCP) for elastoplastic frictional contact problems can be obtained, by substituting eqn (4.3) for $\{\mathrm{d}p_\mathrm{c}\}_i$ and eqn (4.25) for $\{\mathrm{d}\varepsilon^\mathrm{p}\}_i$ into eqn (4.131) and finally both into eqn (4.24), to give:

$$\begin{aligned}
-\{f_\mathrm{p}\}_i &= \{q_\mathrm{p}\}_i + \left[M_\mathrm{pc}\right]_{i-1}\{\mathrm{d}\lambda_\mathrm{c}\}_i + \left[M_\mathrm{pp}\right]_{i-1}\{\mathrm{d}\lambda_\mathrm{p}\}_i\\
&= \{q_\mathrm{p}\}_i + \left[\left[M_\mathrm{pc}\right]_{i-1}\quad\left[M_\mathrm{pp}\right]_{i-1}\right]\begin{Bmatrix}\{\mathrm{d}\lambda_\mathrm{c}\}_i\\ \{\mathrm{d}\lambda_\mathrm{p}\}_i\end{Bmatrix}\\
&= \{q_\mathrm{p}\}_i + \left[\left[M_\mathrm{pc}\right]_{i-1}\quad\left[M_\mathrm{pp}\right]_{i-1}\right]\{\mathrm{d}\lambda\}_i,
\end{aligned}\tag{4.137}$$

in which

$$\begin{aligned}
\{q_\mathrm{p}\}_i &= \{-Y\} + \left[C_\mathrm{p}\right]_{i-1} \cdot \left\{\{\sigma\}_{i-1} + \left[A_\mathrm{c}^\sigma\right]\{[D_\mathrm{c}]\{u_\mathrm{c}\}_{i-1} - \{p_\mathrm{c}\}_{i-1}\}\right\} \\
&\quad + \left[C_\mathrm{p}\right]_{i-1} \cdot \left\{\{F^\sigma\}_i + \left[A_\mathrm{c}^\sigma\right][D_\mathrm{c}]\{E_\mathrm{c}\}_i\right\}, \\
\left[M_\mathrm{pc}\right]_{i-1} &= \left[C_\mathrm{p}\right]_{i-1} \cdot \left[A_\mathrm{c}^\sigma\right][D_\mathrm{c}] \cdot [[B_\mathrm{c}] - [C_\mathrm{c}]], \\
\left[M_\mathrm{pp}\right]_{i-1} &= \left[C_\mathrm{p}\right]_{i-1} \cdot \left[\left[A_\mathrm{c}^\sigma\right][D_\mathrm{c}]\left[B_\mathrm{p}\right]_{i-1} + [R^\sigma]\left[C_\mathrm{p}\right]_{i-1}^\mathrm{T}\right].
\end{aligned} \tag{4.138}$$

In order to determine the unknown variables of the proposed elastoplastic frictional contact problems, it is first required to combine the above two LCPs into a unified form as follows:

$$-\begin{Bmatrix}\{f_\mathrm{c}\}_i \\ \{f_\mathrm{p}\}_i\end{Bmatrix} = \begin{Bmatrix}\{q_\mathrm{c}\}_i \\ \{q_\mathrm{c}\}_i\end{Bmatrix} + \begin{bmatrix}[M_\mathrm{cc}] & [M_\mathrm{cc}]_{i-1} \\ \left[M_\mathrm{pc}\right]_{i-1} & \left[M_\mathrm{pp}\right]_{i-1}\end{bmatrix}\begin{Bmatrix}\{\mathrm{d}\lambda_\mathrm{c}\}_i \\ \{\mathrm{d}\lambda_\mathrm{p}\}_i\end{Bmatrix} \tag{4.139}$$

$$\begin{Bmatrix}((m_\mathrm{c}+1)\cdot C)\times(1) \\ (m_\mathrm{p}\cdot(N_1+N_2))\times(1)\end{Bmatrix} \quad \begin{Bmatrix}((m_\mathrm{c}+1)\cdot C)\times(1) \\ (m_\mathrm{p}\cdot(N_1+N_2))\times(1)\end{Bmatrix}$$

$$\begin{bmatrix}((m_\mathrm{c}+1)\cdot C)\times(1)((m+1)\cdot C) & ((m_\mathrm{c}+1)\cdot C)\times(y\cdot(N_1+N_2)) \\ (m_\mathrm{p}\cdot(N_1+N_2))\times((m+1)\cdot C) & (m_\mathrm{p}\cdot(N_1+N_2))\times(y\cdot(N_1+N_2))\end{bmatrix}$$

$$\begin{Bmatrix}((m_\mathrm{c}+1)\cdot C)\times(1) \\ (m_\mathrm{p}\cdot(N_1+N_2))\times(1)\end{Bmatrix},$$

where the terms in the parentheses, which are defined below, show the number of rows and columns of the corresponding vectors and matrices.

m_c is the number of the orientation of each plane in the piecewise linear approximation of the friction cone, in which values of $m_\mathrm{c} = 2$ in 2D and $m_\mathrm{c} = 8$ in 3D are used in the present study.

C is the number of nodes of the potential contact zone.

m_p is the number of approximating yield functions (planes) or yield modes in the principal stress state, with $m_\mathrm{p} = 6$ for the Tresca and $m_\mathrm{p} = 8$ for the von Mises yield condition in the present study.

N_1, N_2 is the sum of the number of surface (boundary) nodes and interior sample points of the two contacting bodies α ($\alpha = 1, 2$) respectively.

The compact form of eqn (4.139) can then be written:

$$-\{f\}_i = \{q\}_i + [M]_i\{\mathrm{d}\lambda\}_i \tag{4.140}$$

with

$$\begin{matrix}-\{f\}_i \geq 0, \\ \{\mathrm{d}\lambda_i\} \geq 0,\end{matrix} \qquad -\{f\}_i^\mathrm{T}\{\mathrm{d}\lambda_i\} = 0 \qquad \text{(complementarity conditions)}. \tag{4.141}$$

The set of eqns (4.140, 4.141) defines a standard linear complementarity problem whose, general form is (Lemke, 1964, 1968):

$$\{w\} = \{q\} + [M]\{z\}, \tag{4.142a}$$

$$\{w\} \geq 0, \quad \{z\} \geq 0 \qquad \text{(non-negativity conditions)}, \tag{4.142b}$$

$$\{w\}^\mathrm{T}\{z\} = 0 \qquad \text{(complementarity condition)}, \tag{4.142c}$$

where it is required to find the vectors $\{w\}$, $\{z\}$ satisfying eqns (4.142a–c) for a given vector of data $\{q\}$ and a given matrix $[M]$.

Employing Lemke's algorithm, we can solve for $-\{f\}_i$ and $\{d\lambda\}_i$ from eqn (4.140), and subsequently the unknown contact variables, $\{du_c\}_i$ and $\{dp_c\}_i$, can be obtained from eqns (4.133, 4.134) and (4.3), respectively. After this, the displacement components outside the contact surface (eqn (4.116)), as well as the stresses at any internal point and on the surfaces (eqn (4.131)), can be determined for each contacting body. In addition, using the formulation presented in Appendix A, the displacements at any internal point can be found. In fig. 4.5 we present a flowchart describing the incremental solution algorithm for solving the boundary element-complementarity formulation method, based on the particular integrals concept for the elastoplastic contact problems explained in this chapter.

To conclude, a novel procedure has been presented in this chapter for elastoplastic stress analysis in addition to the contact problem analysis. Two- and three-dimensional boundary element formulations for elastoplastic problems based on the particular integral concept (meshless methods) have been derived and transformed into an LCP (Faraji *et al.*, 2003b). On the basis of this LCP, which is an approximate mathematical representation of the elastoplastic constitutive equation, the unknown variables of the elastoplastic problems, such as the stress and deformation (displacement) field, can be determined using Lemke's algorithm. This procedure, which, as far as we know, has not been presented elsewhere, differs from existing iterative or variable stiffness solution algorithms generally employed in elastoplastic problems using the BEM. A summary of this algorithm is presented in Appendix D.

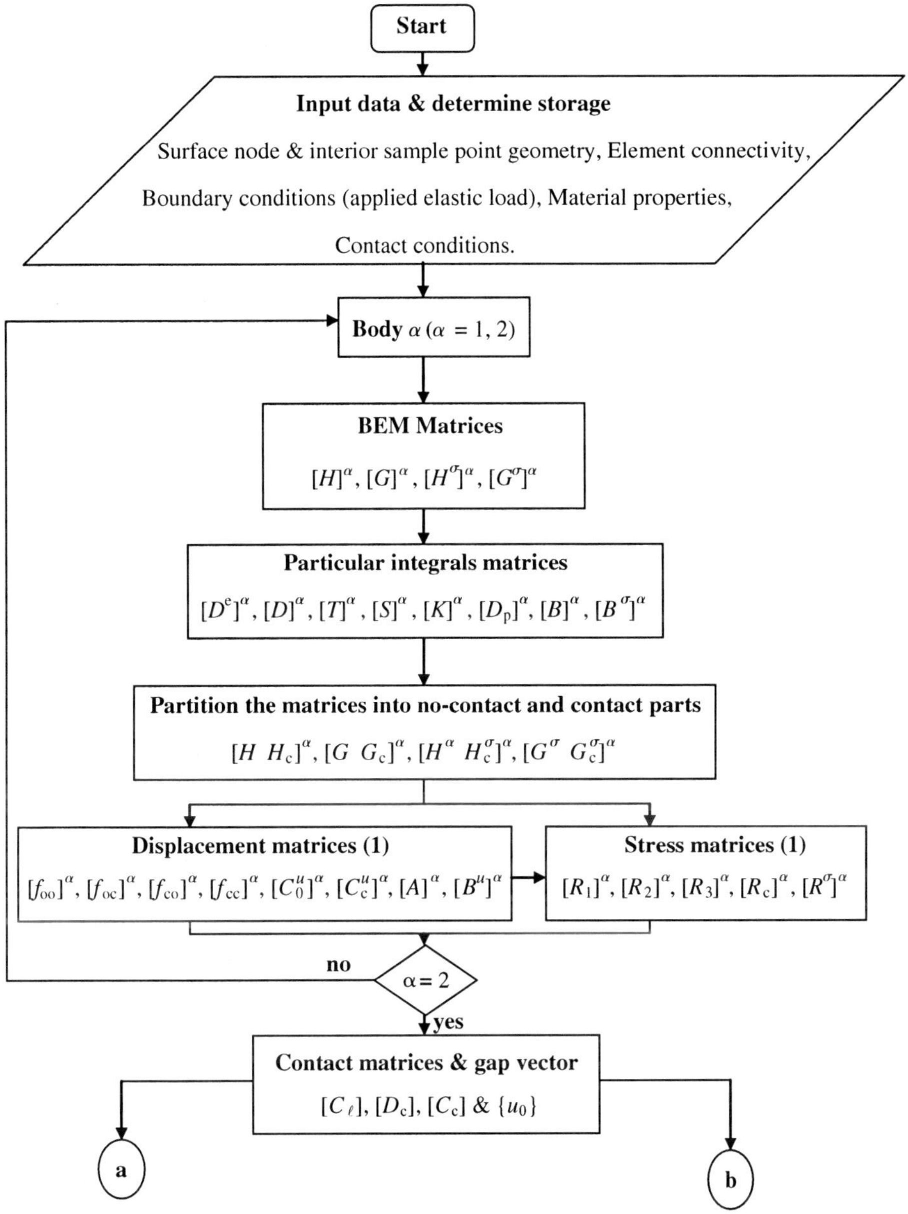

Figure 4.5: Flowchart for elastoplastic contact analysis using BE-MP.

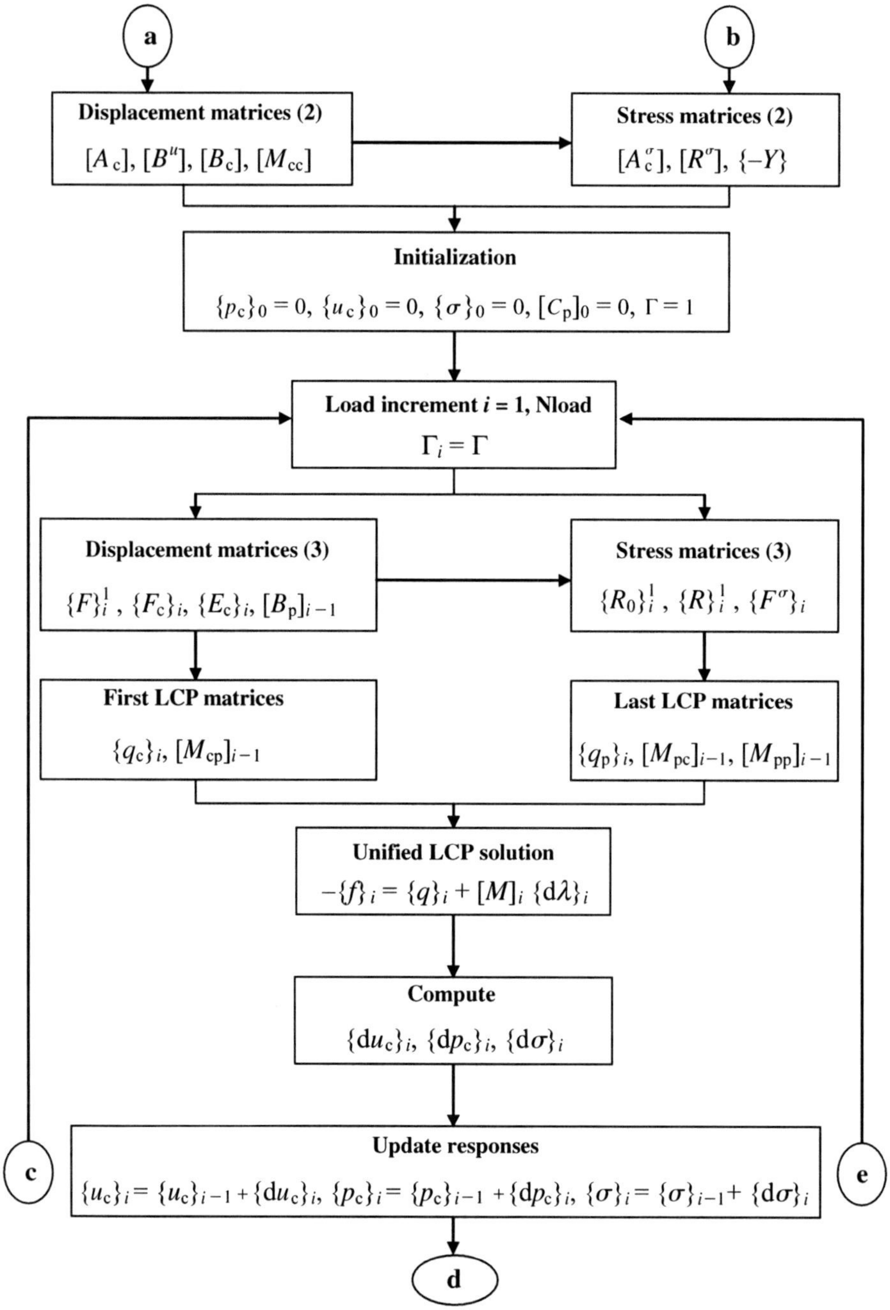

Figure 4.5: (continued).

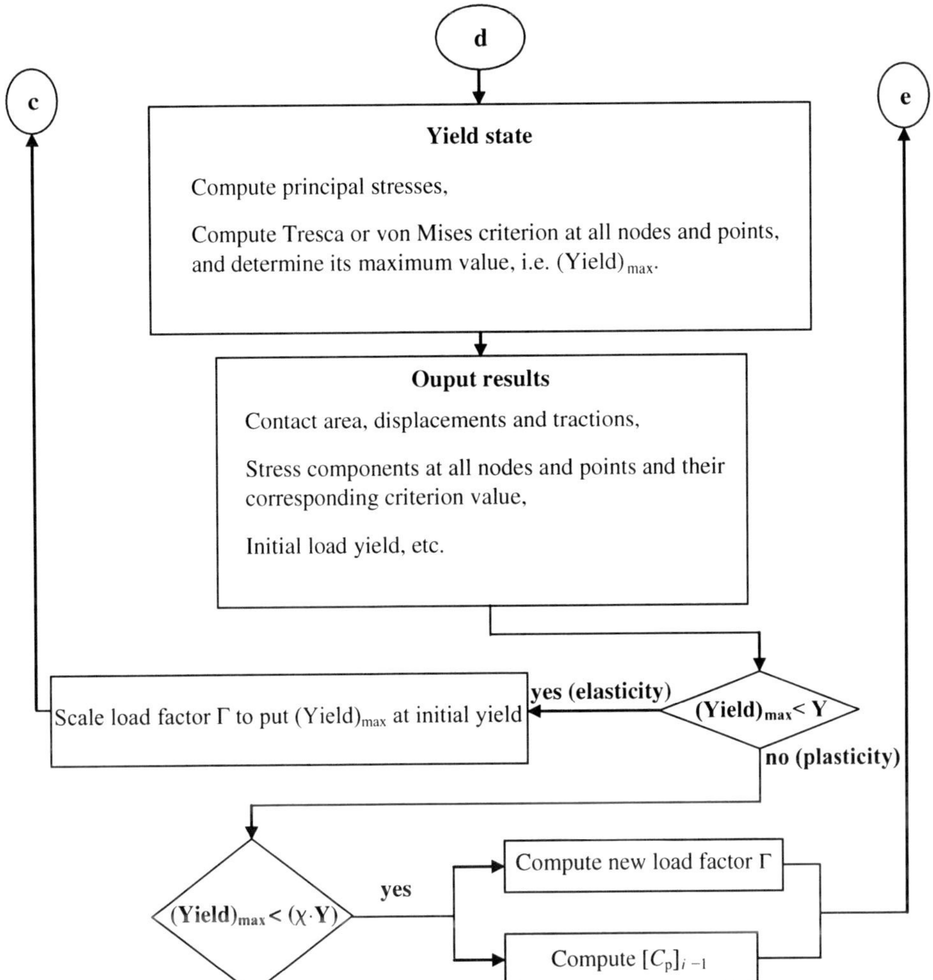

Figure 4.5: (continued).

CHAPTER 5

Three-dimensional elastoplastic frictional contact analysis, Part 2: Applications

5.1 Introduction

In Chapter 4, a general incremental boundary element-mathematical programming method has been formulated to solve elastoplastic frictional contact problems. This provided a unified linear complementarity problem (LCP) which was derived on the basis of a representation of the contact criterion and the ideal elastoplastic constitutive equations as an LCP and of a boundary element formulation based on the use of a particular integrals concepts (meshless method) to eliminate volume integration. Finally, the mathematical programming problem was solved by Lemke's algorithm.

The above analysis has been successfully implemented and added to the package CONTACTA presented in Chapter 3. In the present chapter, some of the 2- and 3D elastoplastic cases of frictional contact problems proposed in the package, are considered. With this addition, the package CONTACTA is now completed and is capable of analyzing frictional contact problems in elastic and/or elastoplastic cases. While there is no limitation to the analysis of elastic and/or elastoplastic contact problems in the 2D case (plane problems), certain limitations also exist on the maximum number of nodes and elements, which can be employed in 3D cases. Nevertheless, it is clear that by updating the package by fully exploiting the new FORTRAN (90, 95) language functionality, certain limitations could be lifted. It should be noted that the largest categories of industrial problems in the nonlinear case are 3D ones. However, many 3D problems can be approximated as 2D cases (plane problems) for reasons of quick evaluation and economy.

The macro-flow diagram of the package CONTACTA for the elastic and elastoplastic cases is shown in fig. 5.1. The main program defines the maximum dimensions of the system of equations and calls three principal modules. The first module, i.e. Input data, calls four block routines: BlCoor, BlBcnd, BlPrel and BlBlem. These routines read the coordinates of the boundary nodes and the interior sample points, boundary conditions, material properties and element node connectivity respectively. If we refer to fig. 4.5, the second module, BETactMp, first forms the BEM matrices using a numerical integration scheme in the cases of singular and non-singular (Stcs and Stns) and the particular integral matrices. Then, it rearranges the matrices according to the boundary conditions, into the displacement and stress matrices in global as well as in local coordinate systems, and subsequently performs the matrix partitioning required by contact problem penalty formulations (elastic or elastoplastic), in order to form the corresponding LCPs. Finally, it calls the Lemke's algorithm (LCP) to calculate the contact variables and present results through the last module (Output). Through this routine, the value of the load at the first yield for which the first node becomes plastic, as well as the stresses and displacements for each node of the contacting bodies, can be obtained and outputted. In order to described the mesh generation of contacting bodies and illustrate the output results, a series of computer programs written in MATLAB (5, 6) language, have also been used.

Using this CONTACTA package, various numerical tests were carried out corresponding to the elastoplastic counterparts of the elastic examples presented in Chapter 3. As before, results are presented in three groups (six examples) in order to illustrate the effectiveness and accuracy of the proposed methodology:

1. Contact between a cylinder and a foundation.
2. Contact between two cylinders (combined normal and tangential loading).
3. Contact between a flat punch and a flat foundation (2D and 3D).

It should be noted that all the numerical results presented in this chapter were obtained by systematically using double precision accuracy in the program, a single value of the penalty parameters taken as $E_n = E_{t1} = E_{t2} = 100E$, where E is Young's modulus, and quadratic isoparametric element approximations (3- and 8-noded). In order to compare our results, labeled

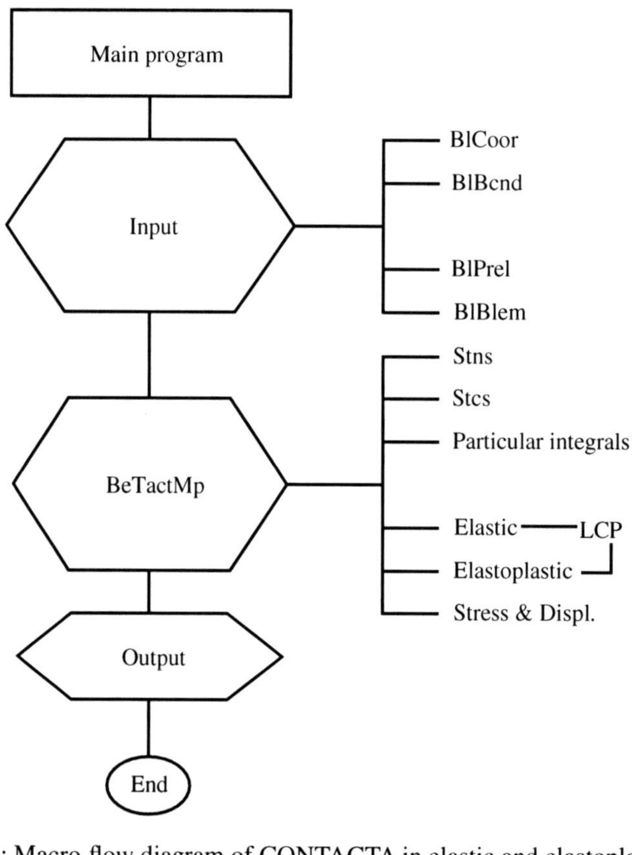

Figure 5.1: Macro-flow diagram of CONTACTA in elastic and elastoplastic cases.

CONTACTA in each figure, with those obtained from other analytical and numerical methods, we had to use the same body geometries, material properties and applied loadings.

(Group 1)

5.2 Contact between an elastoplastic cylinder and a rigid foundation

Assuming a plane strain condition and the von Mises yield criterion, an elastoplastic circular cylinder with radius $R = 60$ mm is pressed on a rigid rectangular cross section foundation with dimensions ($R \times R/2$) under a total normal load of 13,500 N/mm as shown in fig. 5.2. Both bodies have a Poisson's ratio of 0.3, the Young's moduli are 6897 and 10^5 MPa, and the initial

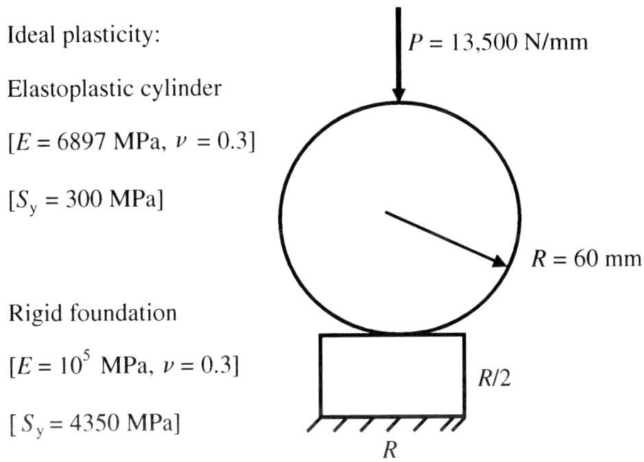

Figure 5.2: Contact between a circular elastoplastic cylinder and a rigid foundation.

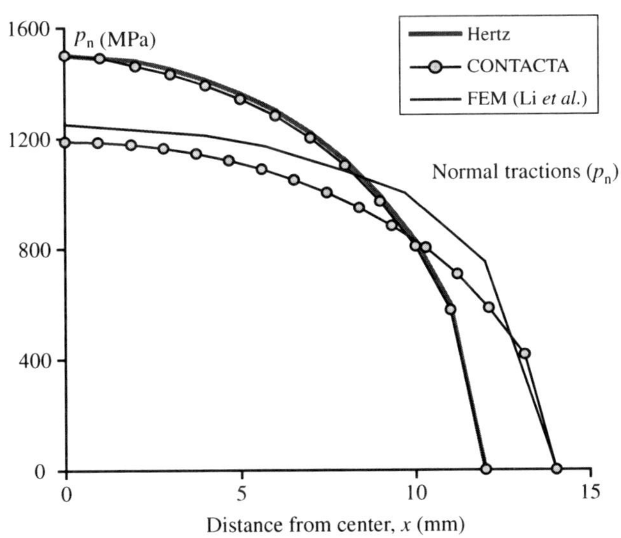

Figure 5.3: Distributions of the normal contact tractions for elastic and elastoplastic cases when $\mu = 0.0$.

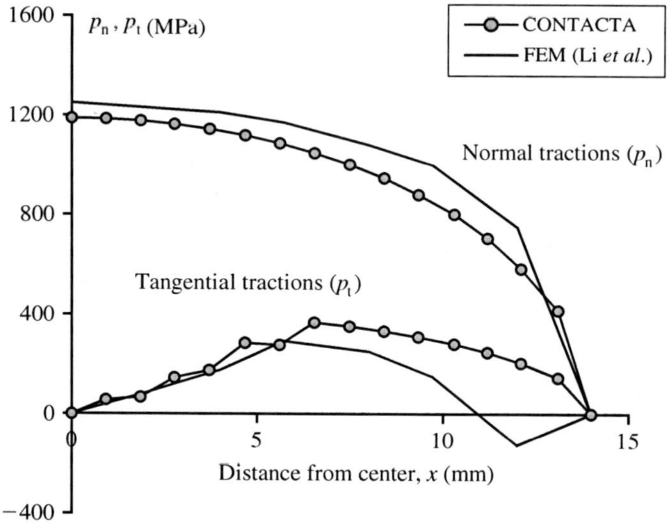

Figure 5.4: Distributions of the normal and tangential contact tractions for the elastoplastic case when $\mu = 0.35$.

yield stresses are 300 and 4350 MPa for the cylinder and the foundation respectively. A soft (non-hardening) material was chosen for the cylinder and a high Young's modulus selected for the foundation to satisfy its rigidity assumption. Both the cylinder and the foundation have been meshed into 48 quadratic elements, with 49 node pairs (24 quadratic elements) in the potential contact zone, and 21 interior sample points within each body near the potential contact zone where plastic deformation is expected.

This 2D elastoplastic contact problem involving linear hardening has been studied using the finite element method (FEM) by Li *et al.* (1997). The distribution of normal traction under both elastic and elastoplastic conditions, with $\mu = 0.0$, are compared in fig. 5.3. When material yielding occurs, the normal traction curve is significantly changed, and is flat compared with the curve for the elastic case. This feature was also reported by Li *et al.* (1997) using the FEM. For the frictional case when $\mu = 0.35$, the normal and tangential tractions versus the length of the contact zone are illustrated in fig. 5.4, together with the results obtained by Li *et al.* (1997). However, these results are not directly comparable because linear hardening was assumed by Li *et al.* (1997), but a similar phenomenon can be observed when the plasticity effect is considered. The comparison of the normal contact tractions in the frictionless and frictional cases shown in figs 5.3 and 5.4, obtained by Li *et al.* (1997) and the present study, indicates that local friction conditions have practically no influence on the normal contact pressure distribution.

(Group 2)

5.3 Contact between two elastoplastic cylinders (combined normal and tangential loading)

The objective here is to study the first critical mode of interwire elastoplastic contact stress analysis of conductors as a plane strain problem by recalling that the contacting wires may be treated as

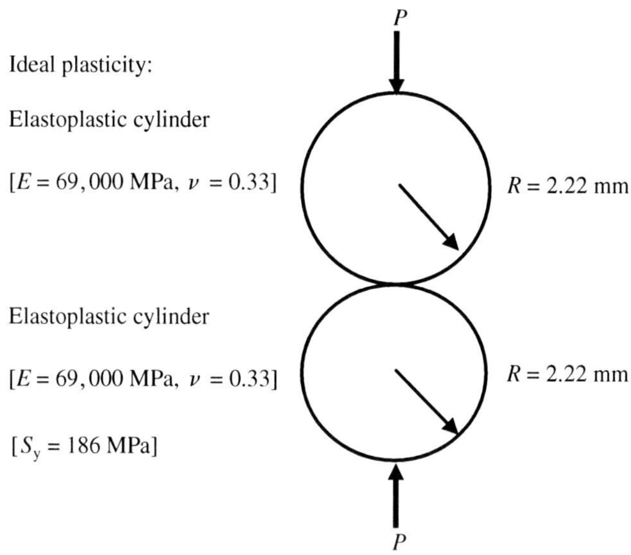

Figure 5.5: Contact between two elastoplastic cylinders with friction.

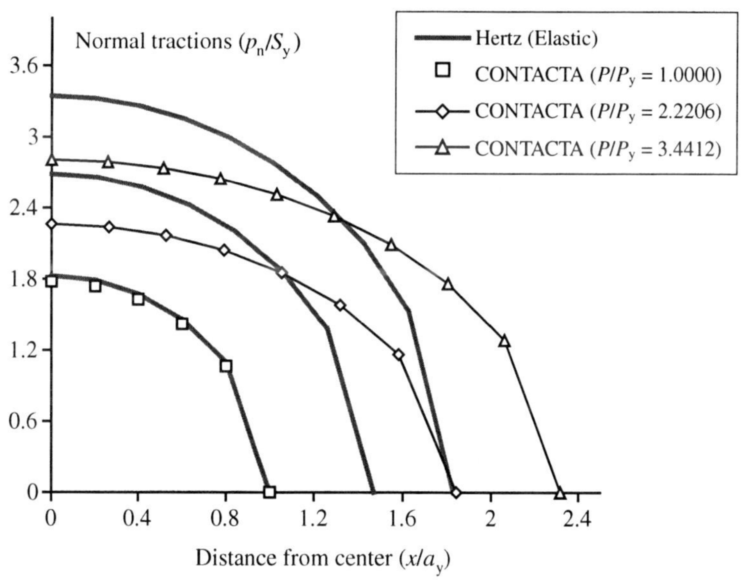

Figure 5.6: Distribution of the normal contact tractions for the elastoplastic case when $\mu = 0.8$.

contacting parallel cylinders. This example is thus aimed at illustrating the influence of combined prescribed normal and tangential loading on the contact status of the two bodies. This apparently is the first numerical solution of this kind of problem, and thus comparative results are not available.

Consider two geometrically identical cylinders, which come into contact under two different loading paths as shown in figs 5.5 and 5.9. Figure 5.5 shows the applied vertical compressive load for which there is no tangential loading. On the other hand, in fig. 5.9 normal and tangential loads

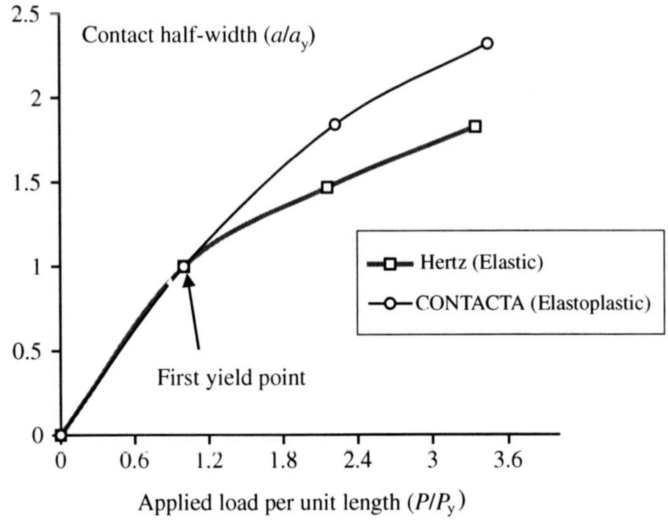

Figure 5 7: Contact half-width-load curve given by Hertz and elastoplastic results for $\mu = 0.8$.

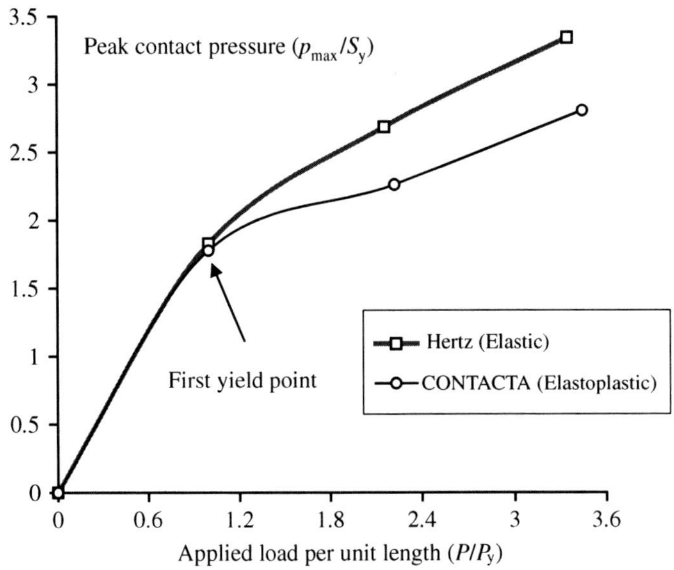

Figure 5.8: Peak contact pressure-load curve given by Hertz and elastoplastic results for $\mu = 0.8$.

are imposed simultaneously. The two cylinders are assumed to be made of aluminum exhibiting elastic–perfectly plastic material behavior with the following practical material properties: Young's modulus 69 GPa, Poisson's ratio 0.33 and yield stress 186 MPa. A local coefficient of friction of 0.8 was chosen for all the results reported in this section. The von Mises yield criterion is used and the plane strain condition is assumed. Both cylinders have been discretized into 48 quadratic elements, with 29 node pairs (14 quadratic elements) in the potential contact zone and 21 interior sample points within each body, near the potential contact zone where plastic deformation is expected.

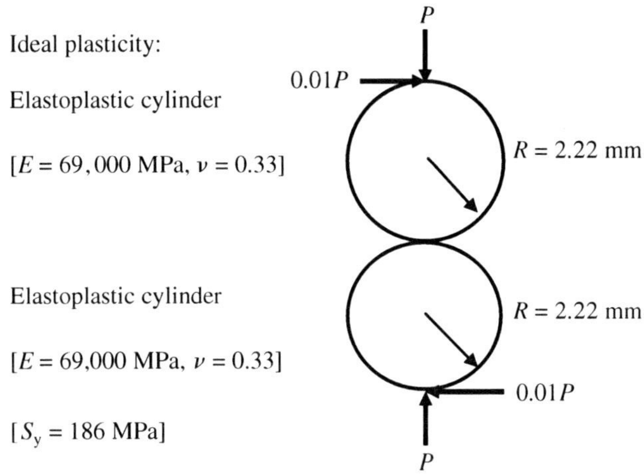

Figure 5.9: Contact between two elastoplastic cylinders with friction.

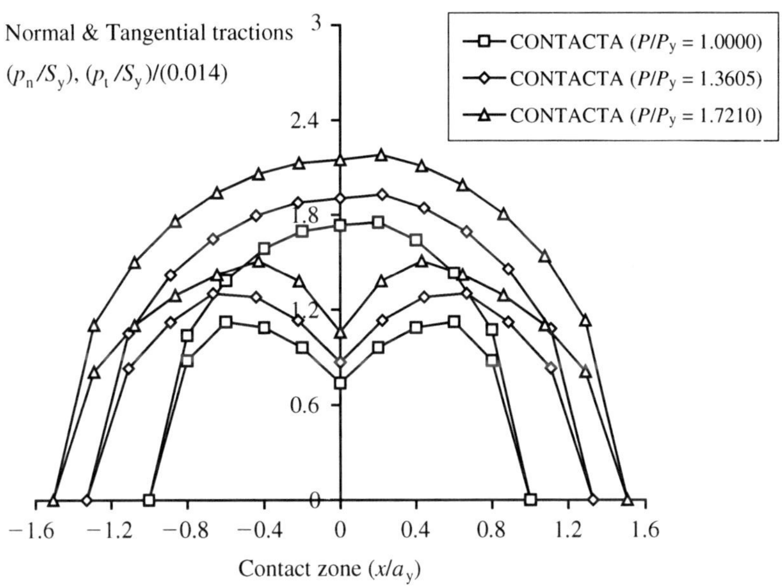

Figure 5.10: Distributions of the normal and tangential contact tractions for the elastoplastic case when $\mu = 0.8$.

The resulting contact normal tractions are shown in fig. 5.6, as a function of location on the contact surface for progressively increasing values of the applied normal load only, following the incremental law given in the flowchart shown in fig. 4.5. The external applied loads, the tractions and the distances from the center of contact zone are normalized with respect to the initial yield load per unit length, P_y, initial yield stress, S_y, and initial yield contact half-width, a_y (the suffix 'y' denotes the point of first yield). Following the von Mises criterion, the analytical load value per unit length at the threshold of plasticity, P_y, when the friction is absent, is 10.4182 N/mm, incipient plastic flow occurs below the contact surface, and the corresponding contact half-width is 0.019501 mm (Johnson, 1985). Will the friction present, the yielding load value per unit length and the contact half-width are found in the present study 10.1180 N/mm and 0.018955 mm, respectively. As seen in fig. 5.6, the contact area increases with a decrease in the maximum normal contact traction. On the other hand, with the load increasing, the pressure distribution although symmetrical, is found to change from an elliptical to a rather flat shape at higher values of the applied normal load. This is due to the plastic behavior of the material beneath the contact zone. On the other hand, in this case, yielding begins at a load of 10.1180 N/mm and starts beneath the surface. The maximum plastic deformation remains inside the cylinders, under the contact surface, and even with the highest loading the boundaries still behave elastically. A normalized comparison of the contact half-width, and the peak contact pressures, versus the applied normal load per unit length, are given in figs 5.7 and 5.8. Results from an elastic as well as an elastoplastic calculation are compared with that obtained using the Hertz theory. It can be seen that, for a specific applied load, the contact half-width is larger and the corresponding maximum contact pressure is smaller than for the elastic solution.

The case of two contacting parallel cylinders simultaneously subjected to an applied normal load per unit length, P, and a small tangential load per unit length, $0.01P$, is shown in fig. 5.9. The normal and tangential tractions in the contact zone, which are normalized, are shown in fig. 5.10. In this case, the resultant yielding load value per unit length is found to be 10.0020 N/mm with a corresponding contact half-width of 0.018750 mm. Comparison of these characteristic values with those obtained in the case of a pure applied normal load, shows that in the presence of the small tangential load, the material yields more readily and yielding still starts below the contact surface. Furthermore, while the tangential tractions are nearly symmetric, the normal tractions are slightly asymmetric about the centerline of the contact zone. Considering the moment about the contact region due to the imposed tangential load, equilibrium in the angular direction requires this slightly non-symmetric pressure distribution.

(Group 3)

5.4 Contact between a flat punch and an elastoplastic flat foundation in 2D

In order to analyze a punch problem, the plane strain elastic–perfectly plastic case is solved assuming the von Mises yield criterion. The example chosen is a flat punch indented into an elastoplastic flat foundation, as shown in fig. 5.11. In this kind of contact problem, the length of the contact zone does not depend on the applied load. Therefore, this is a conforming type of contact problem.

The punch and the foundation are both assumed to have a Poisson's ratio of 0.30, a Young's modulus of 210 GPa and a yield stress of 196 MPa. A uniformly distributed normal traction per unit thickness, q_0, is applied to the upper surface of the punch, while the lower surface of the foundation is held fixed. The coefficient of friction is taken as $\mu = 0.2$. The punch and the foundation boundaries are divided into 42 and 58 quadratic elements respectively, while 41 node pairs (20 quadratic elements) are considered in the potential contact zone. In addition, in the neighborhood of the corners of the punch, which is the region where plastic deformation is expected, 18 and 42 interior sample points are chosen for the punch and the foundation, respectively.

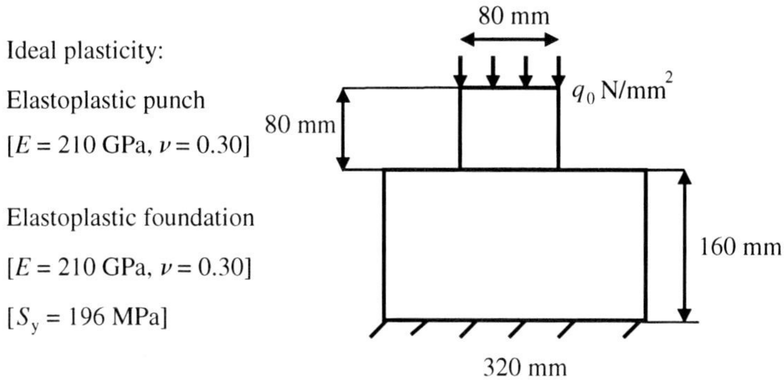

Figure 5.11: Contact between a flat punch and an elastoplastic flat foundation with friction.

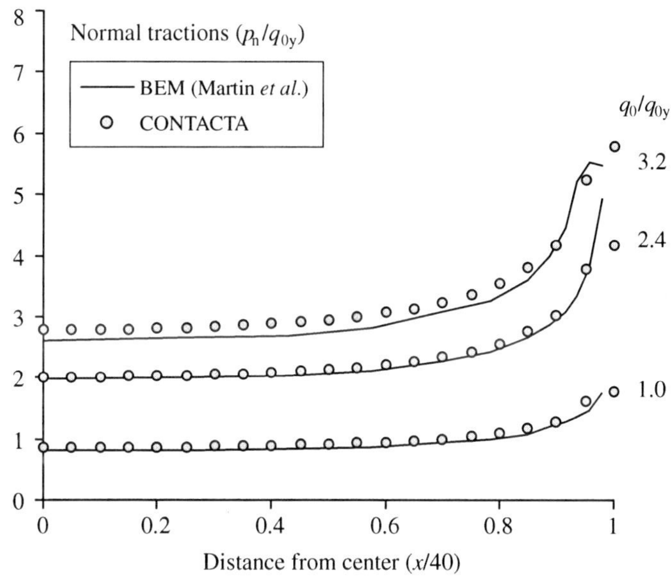

Figure 5.12: Distribution of the normal contact tractions for $\mu = 0.2$.

The first yielding starts at the edges of the punch and spreads towards the interior of the contact zone, and its corresponding load value is $q_{0y} = 51.4$ kN/mm. This phenomenon is the same as that obtained by Martin *et al.* (1994) using the BEM, where the load at first yield was 47.4 kN/mm. Apparently, the explicit form of the stresses produced in an elastic half-space (or half-plane), by the action of an elastic flat punch pressed into the surface in a frictional case, has not yet been evaluated analytically. Nevertheless, the approximate state of stresses within an elastic half-plane in the vicinity of the corners of a rigid flat punch indenting the solid without friction, have been found by Nadai (1963). According to Nadai (1963), the load at first yield can be obtained $q_{0y} = 54.4$ kN/mm. Comparing the yield load values in the frictional case predicted by the present study and by Martin *et al.* (1994), with that calculated based on Nadai's formulations (1963) in the frictionless case, our prediction seems to be more realizable than that of Martin *et al.* (1994).

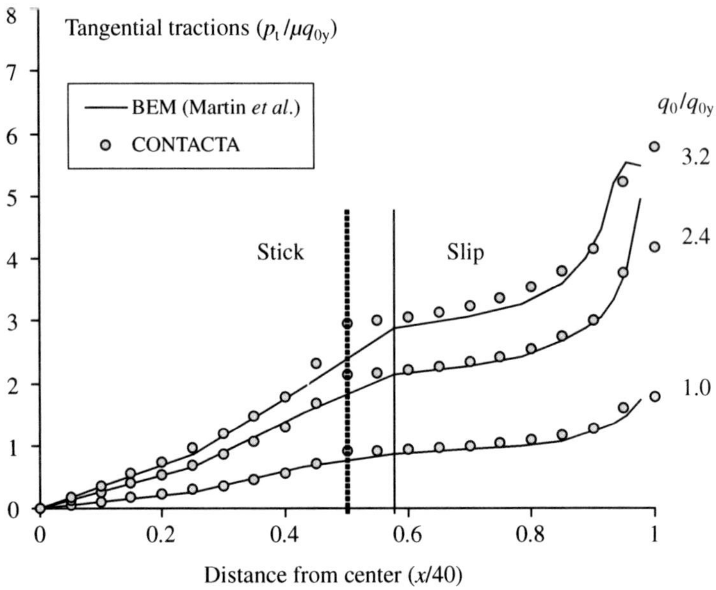

Figure 5.13: Distribution of the tangential contact tractions for $\mu = 0.2$.

The normal and tangential contact tractions normalized with respect to q_{0y} and μq_{0y} are shown in figs 5.12 and 5.13, respectively. The results were compared with those of Martin *et al.* (1994), and found to be in very good agreement. While the present study yields the values of normal and tangential tractions near the end of the contact line, these were not given by Martin *et al.* (1994). The partition of sticking and sliding zones calculated by the two methods is illustrated by the vertical lines in fig. 5.13, and is at 0.50 in the present study and at 0.58 in the work of Martin *et al.* (1994). The problem of partial slip, arising when an elastic half-space (or half-plane) is indented by a rigid flat punch under the action of a purely normal load with a finite coefficient of friction μ between the surfaces, was studied first by Galin (1953) and then more completely by Spence (1973). Spence (1973) has shown that the sticking zone is given by the relationship: $2(1-\nu)\mu/(1-2\nu) = \mathbf{K}(c)/\mathbf{K}(\sqrt{1-c^2})$ (which is the same form as in Galin's result), where $\mathbf{K}(c)$ is the complete elliptic integral of the first kind and the sticking zone, c, is normalized by half the punch width. According to this relationship, the partition can be calculated iteratively to be at 0.45. This indicates that our value seems to be more realizable than that of Martin *et al.* (1994).

5.5 Contact between a flat punch and an elastoplastic flat foundation in 3D

The 3D contact between a flat punch and an elastoplastic flat foundation is shown in fig. 5.14. This was analyzed in the previous example as a plane problem. It is treated here to demonstrate the ability and efficiency of the proposed method in handling 3D problems.

The boundary of the model is discretized using eight-node quadrilateral elements, 28 of which are used for the punch and 52 for the foundation, with four elements defining the contact zone between two bodies. Furthermore, in the vicinity of the corners of the punch, where plastic

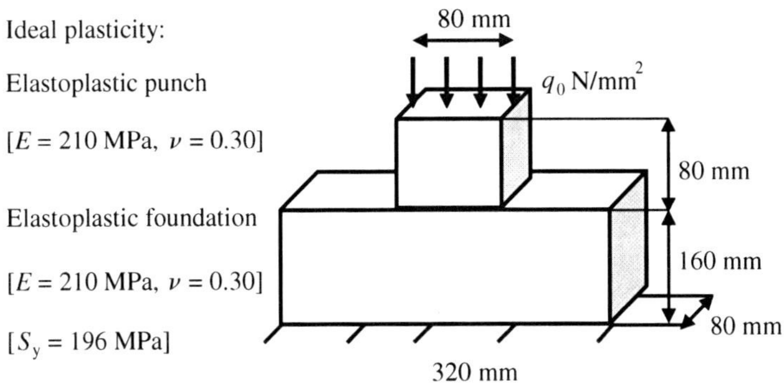

Figure 5.14: Contact between a punch and an elastoplastic foundation with friction in 3D.

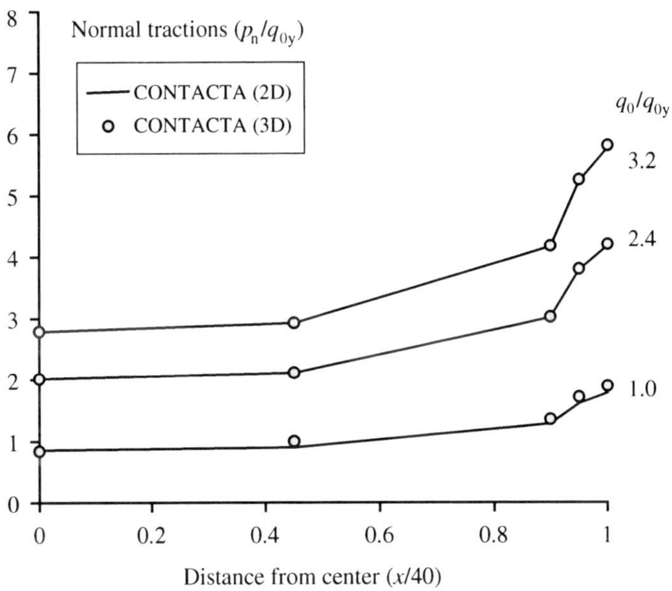

Figure 5.15: Distribution of the normal contact tractions for $\mu = 0.2$.

deformation is expected, 32 and 64 interior sample points are chosen for the punch and the foundation respectively. The punch is loaded by a uniformly distributed normal traction applied at its upper face and the foundation is vertically fixed at the bottom as shown in fig. 5.14. In addition, the geometry of the model and the necessary material property values are illustrated in this figure. The material is assumed to satisfy the von Mises yield criterion as before, and the friction coefficient between the bodies is considered to be $\mu = 0.2$.

Plastic deformation starts at the edges of the punch and spreads over the whole contact area as in 2D. The normal and tangential contact tractions, normalized with respect to q_{0y} and μq_{0y}, are obtained and shown in figs 5.15 and 5.16, respectively. Although no comparative results are available in the literature, numerical results in 3D are compared with those obtained from 2D calculations and are found to be in good agreement. In addition, the boundary between stick and slip zones is obtained and is shown by a vertical line in fig. 5.16.

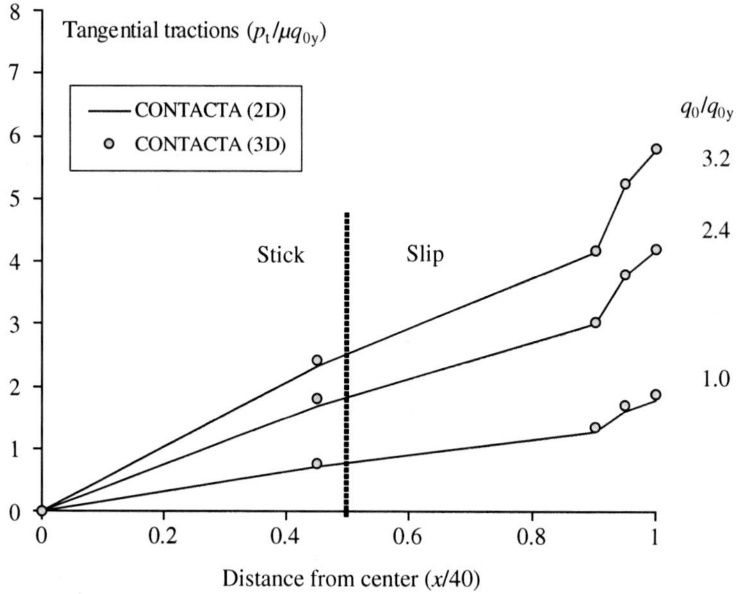

Figure 5.16: Distribution of the tangential contact tractions for $\mu = 0.2$.

CHAPTER 6

Conclusions

A general elastic and elastoplastic analysis method has been presented for the treatment of 2- and 3D contact problems between two deformable bodies undergoing small displacements with or without friction. The normal unilateral and sliding threshold contact laws have been described in the form of normal contact separation (yield) and tangential contact slip (yield) functions following concepts borrowed from elastoplasticity theory. Subsequently, the contact problem has been expressed in a linear complementarity form using a penalty formulation.

In the elastic case, for numerical discretization purposes, the boundary element formulation for the contact problem under applied normal and tangential loads yielded a discrete system of equations relating contact boundary tractions and displacements (contact variables) to the prescribed surface tractions and displacements. These relations have then been coupled with the contact conditions through a penalty formulation. Consequently, the coupled formulation was expressed as a mathematical programming problem, i.e., a linear complementarity problem (LCP) and solved by Lemke's algorithm. This method avoids tedious and costly classical iterative procedures and, if there is no unloading, only one incremental step is required for a solution. Finally, the analysis has been successfully implemented through a package called CONTACTA. The dimension of the contact patch, the normal and tangential tractions distributed in the contact zone, as well as the stresses and displacements for each node of the contacting bodies, have been evaluated using this package. A series of numerical examples, presented in four groups (11 examples), demonstrated the effectiveness and accuracy of the proposed methodology. Very good agreement of the treated problems was obtained with the available analytical and numerical solutions.

In the elastoplastic case, with the material assuming a linear elastic perfectly plastic behavior, a general incremental boundary element-mathematical programming method has been derived to solve elastoplastic frictional contact problems. This unified LCP has been obtained by representing the contact criterion and elastoplastic constitutive equations as an LCP, together with a boundary element formulation using particular integrals (meshless method) to eliminate volume integration. Subsequently, the mathematical programming problem was solved by Lemke's algorithm. This method, which has been derived in a straightforward manner, has a well established mathematical basis, possesses good convergence and eliminates the cumbersome classical trial and error method. The analysis has been successfully implemented through the package CONTACTA. Using this package, where the material is assumed to obey either the von Mises or Tresca criterion with their associated flow rules, the value of the load at the first yield for which the first node becomes plastic, as well as the stress (and displacement) components for each node of the contacting bodies, have been obtained. Furthermore, the contact zone dimension, the distributions of normal and tangential contact tractions and the resulting zone of micro-slip have been determined. In order to indicate the efficiency and accuracy of the proposed methodology, several numerical examples have been treated with the results presented in three groups (six examples). These results are in good agreement with other available analytical and numerical solutions. In particular, the first critical mode of interwire elastoplastic contact stress analysis of conductors has been considered as a plane strain problem, by treating two contacting parallel cylinders under simultaneously applied normal and tangential loads. In the case of the pure applied normal load, excellent agreement was found with analytical values of yield load and the corresponding contact width. The plastic behavior of the material in the contact zone has been found to have a major influence on the pressure distribution between the contacting bodies. In the case of combined normal and tangential loading, it has been demonstrated that the material yields more readily in the presence of a small tangential load.

The methodology and developments presented in this study have been successfully implemented in a series of computer programs (about 100 subroutines), forming a package, called CONTACTA. This package has been written in FORTRAN (66, 77, 90) and MATLAB (5, 6) languages and has been run on a PC computer (Pentium/AMD). Several results illustrated through the present study have been obtained using this package.

Furthermore, the following important and interesting formulas, required in many numerical methods and practical applications, have been explicitly derived:

1. The surface- (boundary-) and interior-points stresses formula for an isotropic elastic body in a compact matrix form (Appendix A).
2. Convenient forms of principal stresses, yield criteria and flow rule in elastoplastic analysis (Appendix C).
3. A novel procedure for 2- and 3D elastoplastic analysis using boundary elements and mathematical programming (Appendix D).

Finally, we can conclude that although our approach seems to be complex, since it involves knowledge from several scientific fields, it is in effect easy to use and can be appropriate in engineering applications. It offers significant promise and potential to handle more complex problems, such as fretting fatigue, large sliding, large deformation, and dynamic applications in both elastic and elastoplastic cases.

APPENDIX A

Displacements and stresses inside a body and stresses on the boundary of a body

A.1 Displacements and stresses inside a body

Following Chapter 2, once the boundary displacements and tractions, $\{u\}$ and $\{t\}$, are known the displacements at any internal point of the body are given by Somiglliania's identity (2.42):

$$u_\ell^i = \int_S u_{\ell k}^* t_k \mathrm{d}S - \int_S t_{\ell k}^* u_k \mathrm{d}S. \tag{A.1}$$

Its discretized form for any internal point i as in eqn (2.51) can be written:

$$\{u\}^i = \sum_{j=1}^{\mathrm{NE}} \left[\int_{S_j} [u]^* \{\phi\} \mathrm{d}s \right] \{t\}^j - \sum_{j=1}^{\mathrm{NE}} \left[\int_{S_j} [t]^* \{\phi\} \mathrm{d}s \right] \{u\}^j , \tag{A.2}$$

where summation $\sum$ from $j = 1$ to NE indicates summations over all the NE elements on the surface S, and S_j is the surface of j element and i is an internal point. Employing a numerical integration scheme such as the Gauss–Legendre quadrature method (Stroud and Secrest, 1966) the above equation may be expressed:

$$\{u\}^i = \sum_{j=1}^{M} \left[G^u\right]^{ij} \{t\}^j - \sum_{j=1}^{M} \left[H^u\right]^{ij} \{u\}^j , \tag{A.3}$$

where M is the number of nodes j on the elements and the terms $[G^u]^{ij}$ and $[H^u]^{ij}$ consist of integrals over the elements to which node j belongs. Because i is an internal point and j is any node on the elements, the integral terms do not contain any singularities and can be easily calculated using the proposed numerical integration method.

Finally, a matrix form of the displacement field can be formulated in terms of the known boundary values:

$$\{u\}^i = \left[G^u\right] \{t\} - \left[H^u\right] \{u\} . \tag{A.4}$$

Note that the superscript i is used to indicate the interior displacement components (quantities).

In order to compute the stresses at any internal point of an isotropic elastic body, one may employ the generalized Hooke's law (eqn (2.35a)):

$$\sigma_{ij} = \frac{E}{(1+\nu)}\left[\frac{\nu}{(1-2\nu)}\delta_{ij}u_{\ell,\ell} + \frac{1}{2}(u_{i,j}+u_{j,i})\right] \quad \text{in } V \text{ and } i, j = 1, 2, 3, \tag{A.5}$$

where

$$u_{\ell,\ell} = u_{1,1} + u_{2,2} + u_{3,3}. \tag{A.6}$$

Differentiating the internal displacements (eqn (A.1)) and substituting into eqn (A.5) yields:

$$\begin{aligned}\sigma_{ij} = &\int_S \left[\frac{E}{(1+\nu)}\left[\frac{\nu}{(1-2\nu)}\delta_{ij}u^*_{\ell k,\ell} + \frac{1}{2}(u^*_{ik,j}+u^*_{jk,i})\right]\right]t_k\mathrm{d}s \\ &- \int_S \left[\frac{E}{(1+\nu)}\left[\frac{\nu}{(1-2\nu)}\delta_{ij}t^*_{\ell k,\ell} + \frac{1}{2}(t^*_{ik,j}+t^*_{jk,i})\right]\right]u_k\mathrm{d}s.\end{aligned} \tag{A.7}$$

Note that the derivatives affect only the fundamental solutions $u^*_{\ell k}$ and $t^*_{\ell k}$, since they are taken at the internal point under consideration. Determining the derivatives, the stress field then takes the form:

$$\sigma_{ij} = \int_S u^*_{kij}t_k\mathrm{d}S - \int_S t^*_{kij}u_k\mathrm{d}S, \tag{A.8}$$

where u^*_{kij} and t^*_{kij} are the third-order fundamental tensors given by:

$$u^*_{kij} = \frac{1}{4A\pi(1-\nu)r^A}\left[(1-2\nu)\left[\delta_{ki}\frac{r_j}{r} + \delta_{kj}\frac{r_i}{r} - \delta_{ij}\frac{r_k}{r}\right] + \frac{Br_ir_jr_k}{r^3}\right], \tag{A.9}$$

$$\begin{aligned}t^*_{kij} = &\frac{G}{2A\pi(1-\nu)r^B}\left[B\frac{\partial r}{\partial n}\left[(1-2\nu)\delta_{ij}\frac{r_k}{r} + \nu\left(\delta_{ik}\frac{r_j}{r} + \delta_{jk}\frac{r_i}{r}\right) - \frac{Cr_ir_jr_k}{r^3}\right]\right. \\ &\left. + B\nu\left(n_i\frac{r_jr_k}{r^2} + n_j\frac{r_ir_k}{r^2}\right) + (1-2\nu)\left(Bn_k\frac{r_ir_j}{r^2} + n_j\delta_{ik} + n_i\delta_{jk}\right) - (1-4\nu)n_k\delta_{ij}\right],\end{aligned} \tag{A.10}$$

where for the 3D case ($A = 2$, $B = 3$, $C = 5$) and for the 2D plane strain case ($A = 1$, $B = 2$, $C = 4$) and, in addition for the 2D plane stress case, E and ν have to be replaced by $E(1 + 2\nu)/(1+\nu)^2$ and $\nu/(1+\nu)$, respectively.

The internal stresses formula (A.8) can be written in a discretized form using the boundary element formulation and corresponding interpolation functions presented in Chapter 2:

$$\sigma^i_{k\ell} = \sum_{j=1}^{NE}\left[\int_{s_j}\langle u^*_{dk\ell}\rangle\{\phi\}\mathrm{d}s\right]\{t\}^j - \sum_{j=1}^{NE}\left[\int_{s_j}\langle t^*_{dk\ell}\rangle\{\phi\}\mathrm{d}s\right]\{u\}^j, \tag{A.11}$$

in which

(a) In 2D

$$\langle u^*_{dk\ell}\rangle = \langle u^*_{1k\ell} \quad u^*_{2k\ell}\rangle, \quad \langle t^*_{dk\ell}\rangle = \langle t^*_{1k\ell} \quad t^*_{2k\ell}\rangle. \tag{A.12a}$$

(b) In 3D

$$\langle u^*_{dk\ell}\rangle = \langle u^*_{1k\ell} \quad u^*_{2k\ell} \quad u^*_{3k\ell}\rangle, \quad \langle t^*_{dk\ell}\rangle = \langle t^*_{1k\ell} \quad t^*_{2k\ell} \quad t^*_{3k\ell}\rangle, \tag{A.12b}$$

where summation $\sum$ from $j = 1$ to NE indicates summations over all the NE elements on the surface S, and S_j is the surface of the j element and i is an internal point.

Applying a numerical integration scheme such as the Gauss–Legendre quadrature method (Stroud and Secrest, 1966), eqn (A.11) can be written without any singularity (such as internal displacements):

$$\sigma_{k\ell}^{i} = \sum_{j=1}^{M} \left\langle G_{\mathrm{d}k\ell}^{\sigma} \right\rangle \{t\}^{j} - \sum_{j=1}^{M} \left\langle H_{\mathrm{d}k\ell}^{\sigma} \right\rangle \{u\}^{j} , \tag{A.13}$$

where M is the number of nodes j on the elements and the terms $\left\langle G_{\mathrm{d}k\ell}^{\sigma} \right\rangle$ and $\left\langle H_{\mathrm{d}k\ell}^{\sigma} \right\rangle$ consist of integrals over the elements to which node j belongs and $\{u\}^{j}$ and $\{t\}^{j}$ are the known displacements and tractions at nodes j.

Finally, the stress field inside the body for any internal point can be written in a compact matrix form:

$$\{\sigma\}^{i} = \left[G^{\sigma}\right]^{i} \{t\} - \left[H^{\sigma}\right]^{i} \{u\} , \tag{A.14}$$

where

(a) In 2D:

$$\{\sigma\}^{i} = \begin{Bmatrix} \sigma_{11} \\ \sigma_{22} \\ \sigma_{12} \end{Bmatrix}^{i} ,$$

$$[G^{\sigma}]^{i} = \begin{bmatrix} G_{111}^{\sigma} & G_{212}^{\sigma} \\ G_{121}^{\sigma} & G_{222}^{\sigma} \\ G_{131}^{\sigma} & G_{232}^{\sigma} \end{bmatrix}^{i} , \tag{A.15}$$

$$[H^{\sigma}]^{i} = \begin{bmatrix} H_{111}^{\sigma} & H_{212}^{\sigma} \\ H_{121}^{\sigma} & H_{222}^{\sigma} \\ H_{131}^{\sigma} & H_{232}^{\sigma} \end{bmatrix}^{i} .$$

(b) In 3D:

$$\{\sigma\}^{i} = \begin{Bmatrix} \sigma_{11} \\ \sigma_{22} \\ \sigma_{33} \\ \sigma_{12} \\ \sigma_{23} \\ \sigma_{31} \end{Bmatrix}^{i} ,$$

$$[G^{\sigma}]^{i} = \begin{bmatrix} G_{111}^{\sigma} & G_{212}^{\sigma} & G_{313}^{\sigma} \\ G_{121}^{\sigma} & G_{222}^{\sigma} & G_{323}^{\sigma} \\ G_{131}^{\sigma} & G_{232}^{\sigma} & G_{333}^{\sigma} \\ G_{141}^{\sigma} & G_{242}^{\sigma} & G_{343}^{\sigma} \\ G_{151}^{\sigma} & G_{252}^{\sigma} & G_{353}^{\sigma} \\ G_{161}^{\sigma} & G_{262}^{\sigma} & G_{363}^{\sigma} \end{bmatrix}^{i} , \tag{A.16}$$

$$[H^{\sigma}]^{i} = \begin{bmatrix} H_{111}^{\sigma} & H_{212}^{\sigma} & H_{313}^{\sigma} \\ H_{121}^{\sigma} & H_{222}^{\sigma} & H_{323}^{\sigma} \\ H_{131}^{\sigma} & H_{232}^{\sigma} & H_{333}^{\sigma} \\ H_{141}^{\sigma} & H_{242}^{\sigma} & H_{343}^{\sigma} \\ H_{151}^{\sigma} & H_{252}^{\sigma} & H_{353}^{\sigma} \\ H_{161}^{\sigma} & H_{262}^{\sigma} & H_{363}^{\sigma} \end{bmatrix}^{i} .$$

Note that the superscript i is used to indicate the interior stress components (quantities).

A.2 Stresses on the boundary of a body

In order to derive the formula for the boundary stresses for an isotropic elastic body, we use the generalized Hooke's law (A.17), the stress–traction relationship (A.19) or Cauchy stress transformation, and the derivative of displacement components given by the interpolation functions (A.20). These are presented in indicial notation for any boundary point s (Rizzo and Shippy, 1968; Kane, 1994):

$$\sigma_{ij} = \frac{E}{(1+\nu)}\left[\frac{\nu}{(1-2\nu)}\delta_{ij}u_{k,k} + \frac{1}{2}(u_{i,j}+u_{j,i})\right] \quad i, j = 1, 2, 3, \tag{A.17}$$

where

$$u_{k,k} = u_{1,1} + u_{2,2} + u_{3,3} \tag{A.18}$$

and

$$t_i = \sigma_{ij}n_j \quad i, j = 1, 2, 3, \tag{A.19}$$

$$\frac{\partial u_i}{\partial a_d} = u_{i,j}\frac{\partial x_j}{\partial a_d} \quad i, j = 1, 2, 3, \tag{A.20}$$

where a_d are the intrinsic coordinates such that in 2D, $a_1 = \xi$, and in 3D, $a_1 = \xi$, $a_2 = \eta$, and x_j are the global coordinates such that $j = 1, 2$ in 2D and $j = 1, 2, 3$ in 3D.

All the values given above are used for the 3D analysis and for the plane strain case ($i, j = 1, 2$). When a plane stress condition is assumed ($i, j = 1, 2$), the same formulation can be maintained simply by replacing E by $E(1+2\nu)/(1+\nu)^2$ and ν by $\nu/(1+\nu)$.

The above set of equations may be written in a compact matrix form:

$$\{\sigma\} = \left[D^{\mathrm{s}}\right]\{\nabla_x u\}, \tag{A.21}$$

$$\{t\} = [\sigma]\{n\}, \tag{A.22}$$

$$\{\nabla_a u\} = \left[u_{i,j}\right]\{\nabla_a x\}, \tag{A.23}$$

where $[\sigma]$ and $[u_{i,j}]$ are the stress tensor and the gradient of the displacement matrix with respect to the global coordinates, respectively, with:

(a) In 2D:

$$\begin{aligned}
\{\sigma\} &= \langle \sigma_{11} \quad \sigma_{22} \quad \sigma_{12}\rangle^{\mathrm{T}}, \\
\{\nabla_x u\} &= \left\langle \frac{\partial u_1}{\partial x_1} \quad \frac{\partial u_1}{\partial x_2} \quad \frac{\partial u_2}{\partial x_1} \quad \frac{\partial u_2}{\partial x_2}\right\rangle^{\mathrm{T}}, \\
\{t\} &= \langle t_1 \quad t_2\rangle^{\mathrm{T}}, \\
\{n\} &= \langle n_1 \quad n_2\rangle^{\mathrm{T}}, \\
\{\nabla_a u\} &= \left\langle \frac{\partial u_1}{\partial \xi} \quad \frac{\partial u_2}{\partial \xi}\right\rangle^{\mathrm{T}}, \\
\{\nabla_a x\} &= \left\langle \frac{\partial x_1}{\partial \xi} \quad \frac{\partial x_2}{\partial \xi}\right\rangle^{\mathrm{T}}.
\end{aligned} \tag{A.24}$$

The matrix $[D^{\mathrm{s}}]$ contains the elastic constants and is given by (in 2D):

$$\left[D^{\mathrm{s}}\right] = \begin{bmatrix} D_1 & 0 & 0 & D_3 \\ D_3 & 0 & 0 & D_1 \\ 0 & D_2 & D_2 & 0 \end{bmatrix}. \tag{A.25}$$

(b) In 3D:

$$\begin{aligned}
\{\sigma\} &= \langle \sigma_{11} \quad \sigma_{22} \quad \sigma_{33} \quad \sigma_{12} \quad \sigma_{23} \quad \sigma_{31} \rangle^{\mathrm{T}}, \\
\{\nabla_x u\} &= \left\langle \frac{\partial u_1}{\partial x_1} \quad \frac{\partial u_1}{\partial x_2} \quad \frac{\partial u_1}{\partial x_3} \quad \frac{\partial u_2}{\partial x_1} \quad \frac{\partial u_2}{\partial x_2} \quad \frac{\partial u_2}{\partial x_3} \quad \frac{\partial u_3}{\partial x_1} \quad \frac{\partial u_3}{\partial x_2} \quad \frac{\partial u_3}{\partial x_3} \right\rangle^{\mathrm{T}}, \\
\{t\} &= \langle t_1 \quad t_2 \quad t_3 \rangle^{\mathrm{T}}, \\
\{n\} &= \langle n_1 \quad n_2 \quad n_3 \rangle^{\mathrm{T}}, \\
\{\nabla_a u\} &= \left\langle \frac{\partial u_1}{\partial \xi} \quad \frac{\partial u_2}{\partial \xi} \quad \frac{\partial u_3}{\partial \xi} \quad \frac{\partial u_1}{\partial \eta} \quad \frac{\partial u_2}{\partial \eta} \quad \frac{\partial u_3}{\partial \eta} \right\rangle^{\mathrm{T}}, \\
\{\nabla_a x\} &= \left\langle \frac{\partial x_1}{\partial \xi} \quad \frac{\partial x_2}{\partial \xi} \quad \frac{\partial x_3}{\partial \xi} \quad \frac{\partial x_1}{\partial \eta} \quad \frac{\partial x_2}{\partial \eta} \quad \frac{\partial x_3}{\partial \eta} \right\rangle^{\mathrm{T}}.
\end{aligned} \tag{A.26}$$

The matrix $[D^{\mathrm{s}}]$ is given (in 3D) by:

$$[D^{\mathrm{s}}] = \begin{bmatrix} D_1 & 0 & 0 & 0 & D_3 & 0 & 0 & 0 & D_3 \\ D_3 & 0 & 0 & 0 & D_1 & 0 & 0 & 0 & D_3 \\ D_3 & 0 & 0 & 0 & D_3 & 0 & 0 & 0 & D_1 \\ 0 & D_2 & 0 & D_2 & 0 & 0 & 0 & 0 & 0 \\ 0 & 0 & 0 & 0 & 0 & D_2 & 0 & D_2 & 0 \\ 0 & 0 & D_2 & 0 & 0 & 0 & D_2 & 0 & 0 \end{bmatrix}, \tag{A.27}$$

in which $D_1 = E(1-\nu)/(1+\nu)(1-2\nu)$, $D_2 = E/2(1+\nu)$ and $D_3 = E\nu/(1+\nu)(1-2\nu)$ are the non-zero elements of matrix $[D^{\mathrm{s}}]$ in 2D as well as in 3D, and the uppercase 'T' indicates a vector or matrix transpose.

In order to obtain a relation for the boundary stresses in terms of tractions and displacements, eqns (A.21–A.23) can be written in the unified form:

$$\begin{bmatrix} [I] & -[D^{\mathrm{s}}] \\ [N] & [0] \\ [0] & [\alpha_{ij}] \end{bmatrix} \begin{Bmatrix} \{\sigma\} \\ \{\nabla_x u\} \end{Bmatrix} = \begin{Bmatrix} \{0\} \\ \{t\} \\ \{\nabla_a u\} \end{Bmatrix}, \tag{A.28}$$

where $[I]$ is the identity matrix of dimensions (3×3) in 2D and (6×6) in 3D, the first $[0]$ is a zero matrix of dimensions (2×4) in 2D and (3×9) in 3D, the last $[0]$ is a zero matrix of dimensions (2×3) in 2D and (6×6) in 3D and finally, $\alpha_{ij} = \partial x_i/\partial a_j$, in which i, j, and a are as defined previously, and also:

(a) In 2D:

$$\begin{aligned}
[N] &= \begin{bmatrix} n_1 & 0 & n_2 \\ 0 & n_2 & n_1 \end{bmatrix}, \\
[\alpha_{ij}] &= \begin{bmatrix} \alpha_{11} & \alpha_{21} & 0 & 0 \\ 0 & 0 & \alpha_{11} & \alpha_{21} \end{bmatrix}.
\end{aligned} \tag{A.29}$$

(b) In 3D:

$$[N] = \begin{bmatrix} n_1 & 0 & 0 & n_2 & 0 & n_3 \\ 0 & n_2 & 0 & n_1 & n_3 & 0 \\ 0 & 0 & n_3 & 0 & n_2 & n_1 \end{bmatrix},$$

$$\left[\alpha_{ij}\right] = \begin{bmatrix} \alpha_{11} & \alpha_{21} & \alpha_{31} & 0 & 0 & 0 & 0 & 0 & 0 \\ 0 & 0 & 0 & \alpha_{11} & \alpha_{21} & \alpha_{31} & 0 & 0 & 0 \\ 0 & 0 & 0 & 0 & 0 & 0 & \alpha_{11} & \alpha_{21} & \alpha_{31} \\ \alpha_{12} & \alpha_{22} & \alpha_{32} & 0 & 0 & 0 & 0 & 0 & 0 \\ 0 & 0 & 0 & \alpha_{12} & \alpha_{22} & \alpha_{32} & 0 & 0 & 0 \\ 0 & 0 & 0 & 0 & 0 & 0 & \alpha_{12} & \alpha_{22} & \alpha_{32} \end{bmatrix}. \tag{A.30}$$

Taking into account the inverse matrix in eqn (A.28), after some manipulation, $\{\sigma\}$ may be obtained as follows:

$$\{\sigma\} = \left[\frac{[D^{s}]}{[\alpha_{ij}0] + [N0][D^{s}]}\right]\left\{\begin{matrix}\{t\} \\ \{\nabla_a u\}\end{matrix}\right\} \equiv \left[[D^{s}1]\;[D^{s}2]\right]\left\{\begin{matrix}\{t\} \\ \{\nabla_a u\}\end{matrix}\right\}$$
$$\equiv \left[D^{s}1\right]\{t\} + \left[D^{s}2\right]\{\nabla_a u\}, \tag{A.31}$$

where

$$\left[\alpha_{ij}0\right] = \begin{bmatrix} [0] \\ \left[\alpha_{ij}\right] \end{bmatrix}, \quad [N0] = \begin{bmatrix} [N] \\ [0] \end{bmatrix}. \tag{A.32}$$

Using the interpolation functions given by eqns (2.49, 2.50) in Chapter 2, to approximate $\{t\}$ and $\{u\}$ in terms of nodal values and substituting into eqn (A.31), the boundary stresses at any boundary point s of an isotropic elastic body can then be written:

$$\{\sigma\}^{s} = \left[D^{s}1\right]\sum_{j=1}^{M}[\phi]\{t\}^{j} + \left[D^{s}2\right]\sum_{j=1}^{M}[\nabla_a\phi]\{u\}^{j}, \tag{A.33}$$

where summation $\sum$ from $j = 1$ to M indicates summations over all the nodes j on the elements on the surface of the body and $\{u\}^{j}$ and $\{t\}^{j}$ are the known displacements and tractions at nodes j.

Finally, the stresses on the boundary of the body can be written in the compact matrix form:

$$\{\sigma\}^{s} = \left[G^{\sigma}\right]^{s}\{t\} + \left[H^{\sigma}\right]^{s}\{u\}. \tag{A.34}$$

Note that the superscript 's' is used to indicate the surface (boundary) stress components (quantities). The relation is the simplest way to determine the stress components at some node point on the surface of the boundary element as an algebraic function instead of an integral equation of the nodal tractions and displacements on that element at that node. This is the way most computer codes used in industry operate, which circumvents the difficulty of singular integration at a boundary point (Rizzo and Shippy, 1968; Kane, 1994).

For further use, it is interesting to assemble eqns (A.14) and (A.34) for both surface- (boundary-) and interior-points stresses:

$$\{\sigma\} = \left\{\begin{matrix}\sigma^{s} \\ \sigma^{i}\end{matrix}\right\} = \begin{bmatrix}[G^{\sigma}]^{s} \\ [G^{\sigma}]^{i}\end{bmatrix}\{t\} - \begin{bmatrix}-[H^{\sigma}]^{s} \\ [H^{\sigma}]^{i}\end{bmatrix}\{u\}$$
$$= \left[G^{\sigma}\right]\{t\} - \left[H^{\sigma}\right]\{u\}. \tag{A.35}$$

Appendix B

Summary of Lemke's algorithm

The linear complementarity problem (LCP) is concerned with solving, by mathematical programming, the following system:

$$\{w\} = \{q\} + [M]\{z\}, \tag{B.1}$$

$$\{w\} \geq 0, \quad \{z\} \geq 0 \quad \text{(non-negativity conditions)}, \tag{B.2}$$

$$\{w\}^{\mathrm{T}}\{z\} = 0 \quad \text{(complementarity condition)}, \tag{B.3}$$

where it is required either to find vectors $\{w\}$ and $\{z\}$ satisfying eqns (B.1–B.3) for a given vector $\{q\}$ and a given matrix $[M]$ or to conclude that no such solution exists. The components of $\{z\}$ are so-called independent variables, while the elements of $\{w\}$ comprise the so-called dependent variables. In the terminology of linear programming, the independent variables are called nonbasic and the dependent variables are called basic. Equation (B.3) indicates that $(\{w\}, \{z\})$ is a pair of complementarity variables. A solution $(\{w\}, \{z\})$ to the above system is called a complementarity feasible solution.

In eqn (B.1), it is clear that if $\{q\} \geq 0$ then we have a trivial problem since $\{w\} = \{q\}, \{z\} = 0$ is a complementarity feasible solution. If there exists at least one $q_{\mathrm{t}} < 0$, then we have an initial basic solution $\{w\} = \{q\}, \{z\} = 0$, which is complementarity but is infeasible. In this case, a non-negative artificial variable z_0 is added to each equation such that $\{w\} = \{q\} + \{z_0\} \geq 0$, choosing a sufficiently large positive value for z_0. Thus, z_0 becomes a basic variable in the first iteration and replaces one of the $\{w\}$ components (say w_r), which becomes nonbasic (i.e., zero) at its introduction. To maintain the complementarity property that $\{w\}\{z\} = 0$, the nonbasic variable z_r is brought into the basic.

A minimum ratio test, similar to that used in the simplex method for solving linear programs, is performed to determine which basic variable has to leave. This algorithm can terminate in two possible ways:

1. Either z_0 leaves the basic vector, which implies that we have a complementarity solution to the LCP (B.1–B.3).
2. Or z_0 may still be in the basic vector and the pivot column may be nonpositive.

When this happens, the algorithm will be unable to solve the LCP. In other words, the minimum ratio test shows that the problem has no solution. This kind of termination is called ray termination.

A detailed analysis of this algorithm can be found in the papers by Lemke (1964, 1968).

APPENDIX C

Convenient forms of principal stresses, yield criteria, and flow rule in elastoplastic analysis

C.1 Introduction

A general method to determine the principal stresses based on some stress invariants is presented. These invariants are employed to define the compact descriptions of the widely used yield criteria, such as those of Tresca and von Mises, which closely approximate metal plasticity behavior. Finally, a general form of the plastic flow rule is derived in the context of the associated theory of plasticity. Particular attention is paid to the treatment of the singularity conditions associated with the Tresca yield criterion. The results given herein in convenient form have been found to be extremely useful in the elastoplastic analysis employed in the present study. These formulations may be considered as alternative forms to those obtained by Nayak and Zienkiewicz (1972); however; the results are identical.

C.2 A general approach to evaluate the principal stresses

In order to evaluate the principal stresses (eigenvalues) of a symmetric stress tensor $[\sigma]$ (Kachanov, 1971; Reddy and Rasmussen, 1982; Batoz and Dhatt, 1990):

$$[\sigma] = \begin{bmatrix} \sigma_{11} & \sigma_{12} & \sigma_{13} \\ & \sigma_{22} & \sigma_{23} \\ \text{Sym.} & & \sigma_{33} \end{bmatrix}, \tag{C.1}$$

the characteristic equation is obtained by setting det $(\sigma_{ij} - \sigma\delta_{ij})$ to zero:

$$\begin{bmatrix} \sigma_{11}-\sigma & \sigma_{12} & \sigma_{13} \\ & \sigma_{22}-\sigma & \sigma_{23} \\ \text{Sym.} & & \sigma_{33}-\sigma \end{bmatrix} = 0 \implies \sigma^3 - I_1\sigma^2 - I_2\sigma - I_3 = 0, \tag{C.2}$$

where I_1, I_2 and I_3 are the invariants associated with the stress tensor $[\sigma]$. The invariants can also be expressed in terms of the components of $[\sigma]$ and/or σ, in which the parameter σ is called a principal stress (eigenvalue):

$$\begin{aligned}
I_1 &= \sigma_{11} + \sigma_{22} + \sigma_{33} = \sigma_1 + \sigma_2 + \sigma_3, \\
I_2 &= -(\sigma_{11}\sigma_{22} + \sigma_{22}\sigma_{33} + \sigma_{33}\sigma_{11}) + \sigma_{12}^2 + \sigma_{23}^2 + \sigma_{31}^2 = -(\sigma_1\sigma_2 + \sigma_2\sigma_3 + \sigma_3\sigma_1), \\
I_3 &= \det[\sigma] = \sigma_{11}\sigma_{22}\sigma_{33} + 2\sigma_{12}\sigma_{23}\sigma_{31} - (\sigma_{11}\sigma_{23}^2 + \sigma_{22}\sigma_{31}^2 + \sigma_{33}\sigma_{12}^2) = \sigma_1\sigma_2\sigma_3.
\end{aligned} \tag{C.3}$$

Finding the roots (the principal stresses) of the cubic eqn (C.2) is not always an easy task. In the alternative method, we seek the principal values (eigenvalues) of the so-called deviatoric stress tensor associated with $[\sigma]$:

$$[\sigma'] = \begin{bmatrix} \sigma'_{11} & \sigma_{12} & \sigma_{13} \\ & \sigma'_{22} & \sigma_{23} \\ \text{Sym.} & & \sigma'_{33} \end{bmatrix} = \begin{bmatrix} \sigma_{11} - \dfrac{I_1}{3} & \sigma_{12} & \sigma_{13} \\ & \sigma_{22} - \dfrac{I_1}{3} & \sigma_{23} \\ \text{Sym.} & & \sigma_{33} - \dfrac{I_1}{3} \end{bmatrix}, \tag{C.4}$$

which has the following associated characteristic equation whose roots are σ':

$$\sigma'^3 - J_2\sigma' - J_3 = 0, \tag{C.5}$$

where

$$\begin{aligned}
J_1 &= \sigma'_{11} + \sigma'_{22} + \sigma'_{33} = \sigma'_1 + \sigma'_2 + \sigma'_3 = 0, \\
J_2 &= \frac{1}{6}\left[(\sigma_{11} - \sigma_{22})^2 + (\sigma_{22} - \sigma_{33})^2 + (\sigma_{33} - \sigma_{11})^2 + 6\left(\sigma_{12}^2 + \sigma_{23}^2 + \sigma_{31}^2\right)\right] \\
&= \frac{1}{6}\left[(\sigma_1 - \sigma_2)^2 + (\sigma_2 - \sigma_3)^2 + (\sigma_3 - \sigma_1)^2\right] \\
&= \frac{1}{2}(\sigma_{11}'^2 + \sigma_{22}'^2 + \sigma_{33}'^2) + \sigma_{12}^2 + \sigma_{23}^2 + \sigma_{31}^2 = I_2 + \frac{I_1^2}{3}, \\
J_3 &= \det[\sigma'] = \sigma'_{11}\sigma'_{22}\sigma'_{33} + 2\sigma_{12}\sigma_{23}\sigma_{31} - (\sigma'_{11}\sigma_{23}^2 + \sigma'_{22}\sigma_{31}^2 + \sigma'_{33}\sigma_{12}^2) \\
&= \sigma_1\sigma_2\sigma_3 = \frac{1}{3}\left(\sigma_1'^3 + \sigma_2'^3 + \sigma_3'^3\right) = I_3 + \frac{1}{3}I_1 I_2 + \frac{2}{27}I_1^3.
\end{aligned} \tag{C.6}$$

The principal stresses (eigenvalues) of the symmetric stress tensor $[\sigma]$ itself can then be computed from:

$$\begin{Bmatrix} \sigma_1 \\ \sigma_2 \\ \sigma_3 \end{Bmatrix} = \begin{Bmatrix} \sigma'_1 \\ \sigma'_2 \\ \sigma'_3 \end{Bmatrix} + \frac{I_1}{3}\begin{Bmatrix} 1 \\ 1 \\ 1 \end{Bmatrix}. \tag{C.7}$$

The cubic equation (C.5) is of a special form that allows a direct computation of its roots. This equation can be solved explicitly by introducing the transformation:

$$\sigma' = 2\sqrt{\frac{J_2}{3}}\cos\omega \tag{C.8}$$

which transforms it into:

$$2\sqrt{\left(\frac{J_2}{3}\right)^3}\left[4\cos^3\omega - 3\cos\omega\right] = J_3. \tag{C.9}$$

The expression in square brackets is equal to cos 3ω. Hence:

$$\cos 3\omega = \frac{J_3}{2}\left(\frac{3}{J_2}\right)^{3/2}. \tag{C.10}$$

If ω_1 is the angle satisfying $0 \leq 3\omega_1 \leq \pi$, whose cosine is given by eqn (C. 10), then $3\omega_1$, $3\omega_1+2\pi$ and $3\omega_1 - 2\pi$ all have the same cosine, and yield three independent roots of eqn (C.5):

$$\begin{Bmatrix} \sigma_1' \\ \sigma_2' \\ \sigma_3' \end{Bmatrix} = 2\sqrt{\frac{J_2}{3}} \begin{Bmatrix} \cos\omega_1 \\ \cos\omega_2 \\ \cos\omega_3 \end{Bmatrix}, \tag{C.11}$$

where

$$\omega_1 = \frac{1}{3}\left[\cos^{-1}\left(\frac{J_3}{2}\left(\frac{3}{J_2}\right)^{3/2}\right)\right], \quad \langle\omega_2 \quad \omega_3\rangle = \left\langle\omega_1 - \frac{2\pi}{3} \quad \omega_1 + \frac{2\pi}{3}\right\rangle, \qquad 0 \leq \omega_1 \leq \pi/3. \tag{C.12}$$

Finally, substituting eqn (C.11) into eqn (C.7), the explicit form of the principal stresses (eigenvalues) of the symmetric stress tensor $[\sigma]$ can be written:

$$\begin{Bmatrix} \sigma_1 \\ \sigma_2 \\ \sigma_3 \end{Bmatrix} = 2\sqrt{\frac{J_2}{3}} \begin{Bmatrix} \cos\omega_1 \\ \cos\omega_2 \\ \cos\omega_3 \end{Bmatrix} + \frac{I_1}{3}\begin{Bmatrix} 1 \\ 1 \\ 1 \end{Bmatrix} \quad \text{with } \sigma_1 \geq \sigma_2 \geq \sigma_3. \tag{C.13}$$

In addition, it will be noted that the eigenvectors of the symmetric stress tensor $[\sigma]$ can be obtained using the relation $(\sigma_{ij} - \sigma\,\delta_{ij})\,e_j = 0$, where $e_j^2 = 1 (j = 1, 2, 3)$.

All the equations given above are used for 3D analysis. When the plane strain or plane stress condition is assumed in a 2D analysis, the same formulation may be maintained by putting $\sigma_{23} = \sigma_{31} = 0$ and $\sigma_{23} = \sigma_{31} = \sigma_{33} = \sigma_3 = \sigma_{33}' = \sigma_3' = 0$ in the plane strain and in plane stress cases, respectively.

It should be noted that the proposed approach is general and may be applied to determine the eigenvalues and eigenvectors of any symmetric matrix whose dimensions are 2×2 or 3×3. An eigenvalue may be a principal stress (such as above), a principal moment of inertia in rigid body mechanics, a time constant in a chemical reaction, a frequency or time constant in a control equation, a principal axis of an ellipse or an ellipsoid, etc.

C.3 Compact descriptions of some yield criteria

It is convenient to express the Tresca and von Mises yield criteria for isotropic materials as functions of two stress invariants (J_2, ω_1) given by eqns (C.6) and (C.12) in the following form $f_p(J_2, \omega_1)$.

1. Tresca yield criterion

This states that yielding begins when the maximum shear stress reaches the value of the maximum shear stress at yielding. It can be written with respect to $\sigma_1 \geq \sigma_2 \geq \sigma_3$:

$$f_p = \sigma_1 - \sigma_3 - Y = 0, \tag{C.14}$$

where Y is the uniaxial yield stress. Substituting for σ_1 and σ_3 from eqn (C.13) into the above equation gives:

$$f_p = 2\sqrt{\frac{J_2}{3}}\left[\cos\omega_1 - \cos\left(\omega_1 + \frac{2\pi}{3}\right)\right] - Y$$
$$= 2\sqrt{J_2}\cos\left(\frac{\pi}{6} - \omega_1\right) - Y = 0. \tag{C.15}$$

2. Von Mises yield criterion

Von Mises suggested that yielding occurs when J_2 reaches a critical value, or:

$$f_p = \sqrt{3J_2} - Y = 0. \tag{C.16}$$

The Tresca regular hexagonal cylinder as well as the von Mises circular cylinder can be illustrated in the principal stress space as follows:

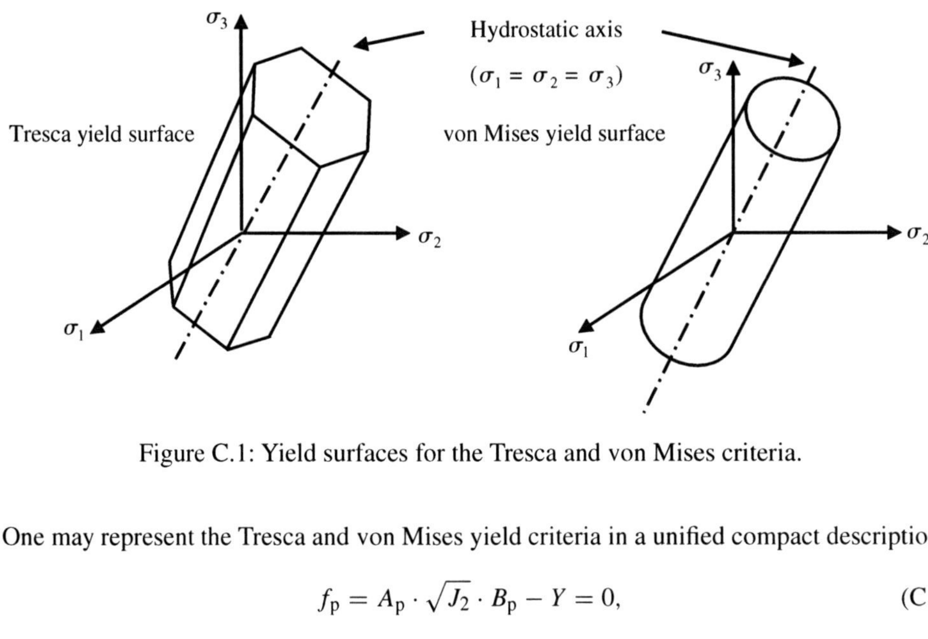

Figure C.1: Yield surfaces for the Tresca and von Mises criteria.

One may represent the Tresca and von Mises yield criteria in a unified compact description:

$$f_p = A_p \cdot \sqrt{J_2} \cdot B_p - Y = 0, \tag{C.17}$$

in which $A_p = 2$ and $B_p = \cos(\pi/6 - \omega_1)$ for the Tresca criterion, and $A_p = \sqrt{3}$ and $B_p = 1$ for the von Mises criterion.

C.4 Plastic flow rule

In the associated theory of plasticity, the plastic strain increment is given by (Kachanov, 1971; Martin, 1975; Lemaitre and Chaboche, 1988; Hill, 1983; Chen and Saleeb, 1994):

$$d\varepsilon_{ij}^p = \frac{\partial f_p}{\partial \sigma_{ij}} d\lambda_p, \tag{C.18}$$

where $\mathrm{d}\lambda_\mathrm{p}$ is a proportionality factor, termed the plastic multiplier, and f_p is the yield surface. The plastic flow rule, i.e. plastic stress–strain relation (C.18) is termed the normality condition, since $\partial f_\mathrm{p}/\partial\sigma_{ij}$ is a vector in the six-dimensional stress space $\sigma_{ij}(i, j = 1, 2, 3)$, which is directed along the normal to the yield surface at the stress point under consideration.

Substituting the unified compact description of the yield criteria (C.17) into the above equation, after simplification, one can represent the flow rule in a general form:

$$\mathrm{d}\varepsilon_{ij}^\mathrm{p} = \left[C_2 \cdot \frac{\partial J_2}{\partial\sigma_{ij}} + C_3 \cdot \frac{\partial J_3}{\partial\sigma_{ij}}\right]\mathrm{d}\lambda_\mathrm{p}. \tag{C.19}$$

This can be explicitly written:

(a) In 2D:

$$\begin{Bmatrix} \mathrm{d}\varepsilon_{11}^\mathrm{p} \\ \mathrm{d}\varepsilon_{22}^\mathrm{p} \\ \mathrm{d}\varepsilon_{12}^\mathrm{p} \end{Bmatrix} = \left[C_2 \cdot \begin{Bmatrix} \dfrac{\partial J_2}{\partial\sigma_{11}} \\ \dfrac{\partial J_2}{\partial\sigma_{22}} \\ \dfrac{\partial J_2}{\partial\sigma_{12}} \end{Bmatrix} + C_3 \cdot \begin{Bmatrix} \dfrac{\partial J_3}{\partial\sigma_{11}} \\ \dfrac{\partial J_3}{\partial\sigma_{22}} \\ \dfrac{\partial J_3}{\partial\sigma_{12}} \end{Bmatrix} \right] \mathrm{d}\lambda_\mathrm{p}, \tag{C.20a}$$

$$\left\langle \frac{\partial J_2}{\partial\sigma_{11}} \quad \frac{\partial J_2}{\partial\sigma_{22}} \quad \frac{\partial J_2}{\partial\sigma_{12}} \right\rangle = \left\langle \sigma'_{11} \quad \sigma'_{22} \quad 2\sigma_{12} \right\rangle, \tag{C.20b}$$

$$\left\langle \frac{\partial J_3}{\partial\sigma_{11}} \quad \frac{\partial J_3}{\partial\sigma_{22}} \quad \frac{\partial J_3}{\partial\sigma_{12}} \right\rangle = \left\langle \sigma'_{22}\sigma'_{33} \quad \sigma'_{33}\sigma'_{11} \quad 2\sigma'_{33}\sigma_{12} \right\rangle + \frac{1}{3} J_2 \left\langle 1 \quad 1 \quad 0 \right\rangle. \tag{C.20c}$$

(b) In 3D:

$$\begin{Bmatrix} \mathrm{d}\varepsilon_{11}^\mathrm{p} \\ \mathrm{d}\varepsilon_{22}^\mathrm{p} \\ \mathrm{d}\varepsilon_{33}^\mathrm{p} \\ \mathrm{d}\varepsilon_{12}^\mathrm{p} \\ \mathrm{d}\varepsilon_{23}^\mathrm{p} \\ \mathrm{d}\varepsilon_{31}^\mathrm{p} \end{Bmatrix} = \left[C_2 \cdot \begin{Bmatrix} \dfrac{\partial J_2}{\partial\sigma_{11}} \\ \dfrac{\partial J_2}{\partial\sigma_{22}} \\ \dfrac{\partial J_2}{\partial\sigma_{33}} \\ \dfrac{\partial J_2}{\partial\sigma_{12}} \\ \dfrac{\partial J_2}{\partial\sigma_{23}} \\ \dfrac{\partial J_2}{\partial\sigma_{31}} \end{Bmatrix} + C_3 \cdot \begin{Bmatrix} \dfrac{\partial J_3}{\partial\sigma_{11}} \\ \dfrac{\partial J_3}{\partial\sigma_{22}} \\ \dfrac{\partial J_3}{\partial\sigma_{33}} \\ \dfrac{\partial J_3}{\partial\sigma_{12}} \\ \dfrac{\partial J_3}{\partial\sigma_{23}} \\ \dfrac{\partial J_3}{\partial\sigma_{31}} \end{Bmatrix} \right] \mathrm{d}\lambda_\mathrm{p}, \tag{C.21a}$$

$$\begin{aligned} &\left\langle \frac{\partial J_2}{\partial\sigma_{11}} \quad \frac{\partial J_2}{\partial\sigma_{22}} \quad \frac{\partial J_2}{\partial\sigma_{33}} \quad \frac{\partial J_2}{\partial\sigma_{12}} \quad \frac{\partial J_2}{\partial\sigma_{23}} \quad \frac{\partial J_2}{\partial\sigma_{31}} \right\rangle \\ &= \left\langle \sigma'_{11} \quad \sigma'_{22} \quad \sigma'_{33} \quad 2\sigma_{12} \quad 2\sigma_{23} \quad 2\sigma_{31} \right\rangle, \end{aligned} \tag{C.21b}$$

$$\left\langle \frac{\partial J_3}{\partial \sigma_{11}} \quad \frac{\partial J_3}{\partial \sigma_{22}} \quad \frac{\partial J_3}{\partial \sigma_{33}} \quad \frac{\partial J_3}{\partial \sigma_{12}} \quad \frac{\partial J_3}{\partial \sigma_{23}} \quad \frac{\partial J_3}{\partial \sigma_{31}} \right\rangle$$

$$= \Big\langle \sigma'_{22}\sigma'_{33} - \sigma_{23}^2 \quad \sigma'_{33}\sigma'_{11} - \sigma_{31}^2 \quad \sigma'_{11}\sigma'_{22} - \sigma_{12}^2 \quad 2(\sigma_{23}\sigma_{31} - \sigma'_{33}\sigma_{12}) \quad 2(\sigma_{31}\sigma_{12} - \sigma'_{11}\sigma_{23}) \quad 2(\sigma_{12}\sigma_{23} - \sigma'_{22}\sigma_{31}) \Big\rangle$$

$$+ \frac{1}{3} J_2 \langle 1 \quad 1 \quad 1 \quad 0 \quad 0 \quad 0 \rangle, \tag{C.21c}$$

where on the basis of the Tresca yield criterion:

$$C_3 = -\frac{\sqrt{3}}{J_2} \cdot \frac{\sin\left(\frac{\pi}{6} - \omega_1\right)}{\sin 3\omega_1}, \quad C_2 = \frac{\cos\left(\frac{\pi}{6} - \omega_1\right)}{\sqrt{J_2}} - \frac{3J_3}{2J_2} \cdot C_3, \tag{C.22a}$$

and for the von Mises yield criterion:

$$C_3 = 0, \quad C_2 = \frac{1}{2}\sqrt{\frac{3}{J_2}}. \tag{C.22b}$$

It should be noted that, as seen from eqn (C.22a), when ω_1 approaches 0 and $\pi/3$, C_2 and C_3 become indeterminate. This means that there are some singularity conditions, which only arise at the corner of the Tresca yield criterion, where the direction of the plastic strain increment vector is not uniquely determined. In order to overcome this difficulty, one can return to the original eqn (C.15) and rewrite this in terms of the explicit values $\omega_1 = 0$ and $\pi/3$, giving:

$$f_\mathrm{p} = \sqrt{3J_2} - Y = 0. \tag{C.23}$$

This equation is the von Mises criterion, which physically states that the direction of the plastic strain increment vector at the corners of the Tresca criterion is equivalent to that given by the von Mises circle, which also passes through the corner (fig. C.1). For numerical computation, one can use the general eqn (C.22a) for both C_2 and C_3 for all values of $1° \leq \omega_1 \leq 59°$ and only employ eqn (C.22b) outside this interval.

APPENDIX D

Summary of three-dimensional elastoplastic analysis using boundary elements and mathematical programming

In Chapter 4, by introducing the plastic yield function as well as the associated plastic flow rule for the linear elastic perfectly plastic behavior of the material, the elastoplastic constitutive equation is expressed in the form of a linear complementarity problem (LCP). On the basis of this LCP and of a boundary element formulation that utilizes the concept of the particular integrals (meshless method) to avoid volume integration, an incremental boundary element–complementarity method to solve elastoplastic problems is summarized in this appendix. Finally, the associated mathematical programming problem can be solved by using Lemke's algorithm to obtain the unknown variables of the problem (Faraji *et al.*, 2003b). The reader can also refer to the papers of Maier and Novati (1983) who used the BEM as well as to those of Zhong and Zhang (1988), Sha *et al.* (1990), and Zhu and Jin (1993) who employed the FEM to solve problems in plasticity.

In Chapter 4, the LCP of the constitutive laws of ideal plasticity in matrix form is presented at the ith incremental step:

$$\{f_{\mathrm{p}}\}_i = \{\{-Y\} + [C_{\mathrm{p}}]_{i-1}\{\sigma\}_{i-1}\} + [C_{\mathrm{p}}]_{i-1}\{\mathrm{d}\sigma\}_i, \tag{D.1}$$

$$\{\mathrm{d}\varepsilon^{\mathrm{p}}\}_i = [C_{\mathrm{p}}]_{i-1}^{\mathrm{T}}\{\mathrm{d}\lambda_{\mathrm{p}}\}_i, \tag{D.2}$$

$$\begin{matrix}\{f_{\mathrm{p}}\}_i \le 0, \\ \{\mathrm{d}\lambda_{\mathrm{p}}\}_i \ge 0,\end{matrix} \quad \{f_{\mathrm{p}}\}_i^{\mathrm{T}}\{\mathrm{d}\lambda_{\mathrm{p}}\}_i = 0, \tag{D.3}$$

where $[C_{\mathrm{p}}]_{i-1}$ is given in Section 4.4 for the Tresca and von Mises criteria.

We also recall that the boundary element formulation for elastoplastic problems, based on the concept of particular integrals (solutions), in terms of the incremental initial strain, was given by:

$$[H]\{\mathrm{d}u\} = [G]\{\mathrm{d}t\} + [B][D_{\mathrm{p}}]\{\mathrm{d}\varepsilon^{\mathrm{p}}\}, \tag{D.4}$$

$$\{\mathrm{d}\sigma\} = [G^{\sigma}]\{\mathrm{d}t\} - [H^{\sigma}]\{\mathrm{d}u\} + [B^{\sigma}][D_{\mathrm{p}}]\{\mathrm{d}\varepsilon^{\mathrm{p}}\}. \tag{D.5}$$

By applying the boundary conditions, i.e., eqns (4.48a,b), with respect to $\{\mathrm{d}u\}$ and $\{\Gamma\mathrm{d}\bar{t}\}$ that represent the boundary unknowns and knowns, and rearranging the corresponding columns and rows in the matrices $[H]$ and $[G]$ as well as those of $[H^{\sigma}]$ and $[G^{\sigma}]$, one finds at the ith incremental step:

$$[H]\{du\}_i = [G]\{\Gamma d\bar{t}\}_i + [B][D_{\text{p}}]\{d\varepsilon^{\text{p}}\}_i, \tag{D.6}$$

$$\{d\sigma\}_i = [G^{\sigma}]\{\Gamma d\bar{t}\}_i - [H^{\sigma}]\{du\}_i + [B^{\sigma}][D_{\text{p}}]\{d\varepsilon^{\text{p}}\}_i. \tag{D.7}$$

Working with eqn (D.6) and multiplying the inverse of matrix $[H]$ with matrices $[G]$ and $[[B][D_p]]$, these equations become:

$$\begin{aligned}\{du\}_i &= \lfloor [H]^{-1}[G]\rfloor \{\Gamma d\bar{t}\}_i + \lfloor [H]^{-1}[B][D_{\text{p}}]\rfloor \{d\varepsilon^{\text{p}}\}_i \\ &= [f_{\text{oo}}]\{\Gamma d\bar{t}\}_i + \left[C_{\text{o}}^{u}\right]\{d\varepsilon^{\text{p}}\}_i,\end{aligned} \tag{D.8}$$

where $[f_{\text{oo}}]$ is called the flexibility matrix.

Substituting the above equation into eqn (D.7), after some manipulations, yields:

$$\{d\sigma\}_i = \{F^{\sigma}\}_i + [R^{\sigma}]\{d\varepsilon^{\text{p}}\}_i, \tag{D.9}$$

where

$$\{F^{\sigma}\}_i = [G^{\sigma} - H^{\sigma} f_{\text{oo}}]\{\Gamma d\bar{t}\}_i, \tag{D.10}$$

$$\{R^{\sigma}\} = [-H^{\sigma} C_{\text{o}}^{u} + B^{\sigma} D_{\text{p}}]. \tag{D.11}$$

Finally, the linear complementarity problem (LCP) for elastoplastic analysis may be obtained, by substitution of eqn (D.2) for $\{d\varepsilon^{\text{p}}\}_i$ into eqn (D.9), and then by substituting the resulting expression into eqn (D.1) after taking account of eqn (D.3):

$$\begin{aligned}&-\{f_{\text{p}}\}_i = \{q_{\text{p}}\}_i + \lfloor M_{\text{pp}}\rfloor_{i-1}\{d\lambda_{\text{p}}\}_i \\ &\{(m_{\text{p}}\cdot N)\times(1)\}\ \{(m_{\text{p}}\cdot N)\times(1)\}\ [(m_{\text{p}}\cdot N)\times(m_{\text{p}}\cdot N)]\ \{(m_{\text{p}}\cdot N)\times(1)\},\end{aligned} \tag{D.12}$$

in which

$$\begin{aligned}\{q_{\text{p}}\}_i &= \{-Y\} + [C_{\text{p}}]_{i-1}\cdot\{\{\sigma\}_{i-1} + \{F^{\sigma}\}_i\}, \\ [M_{\text{pp}}]_{i-1} &= [C_{\text{p}}]_{i-1}\cdot[[R^{\sigma}][C_{\text{p}}]_{i-1}^{\text{T}}]\end{aligned} \tag{D.13}$$

with

$$\begin{matrix}-\{f_{\text{p}}\}_i \le 0, \\ \{d\lambda_{\text{p}}\}_i \ge 0,\end{matrix} \quad -\{f_{\text{p}}\}_i^{\text{T}}\{d\lambda_{\text{p}}\}_i = 0. \tag{D.14}$$

Notice that the terms in the parentheses under eqn (D.12) indicate the number of rows and columns of the corresponding vectors and matrices and are defined as follows:

m_{p} is the number of approximating yield functions (planes) or yield modes in the principal stress space, which is assumed equal to 6 for the Tresca yield condition, and to 8 for the von Mises yield condition in the present study.

N is the sum of the number of surface (boundary) nodes and interior sampling points of the problem.

The set of eqns (D.12–D.14) defines a standard linear complementarity problem whose general form can be written:

$$\{w\} = \{q\} + [M]\{z\}, \tag{D.15a}$$

$$\{w\} \ge 0,\ \ \{z\} \ge 0 \quad \text{(non-negativity conditions)}, \tag{D.15b}$$

$$\{w\}^{\text{T}}\{z\} = 0 \quad \text{(complementarity condition)}, \tag{D.15c}$$

where it is required to find vectors $\{w\}$, $\{z\}$ satisfying eqns (4.15a–c) for a given vector $\{q\}$ and a given matrix $[M]$.

By making use of Lemke's algorithm, $-\{f_{\mathrm{p}}\}_i$ and $\{\mathrm{d}\lambda_{\mathrm{p}}\}_i$ may be solved from eqn (D.12) and consequently the unknown boundary displacement and tractions components of the problem, $\{\mathrm{d}u\}$, can be obtained by eqn (D.8) using eqn (D.2). Subsequently, the stresses at any internal point and stresses on the surfaces eqn (D.9), can be determined. Moreover, using the formulation presented in Appendix A, the displacements at any internal point can be found.

The flowchart in fig. D.1 depicts the algorithm of the incremental boundary element-complementary method, based on the particular integrals concept for solving the elastoplastic problems described in this appendix.

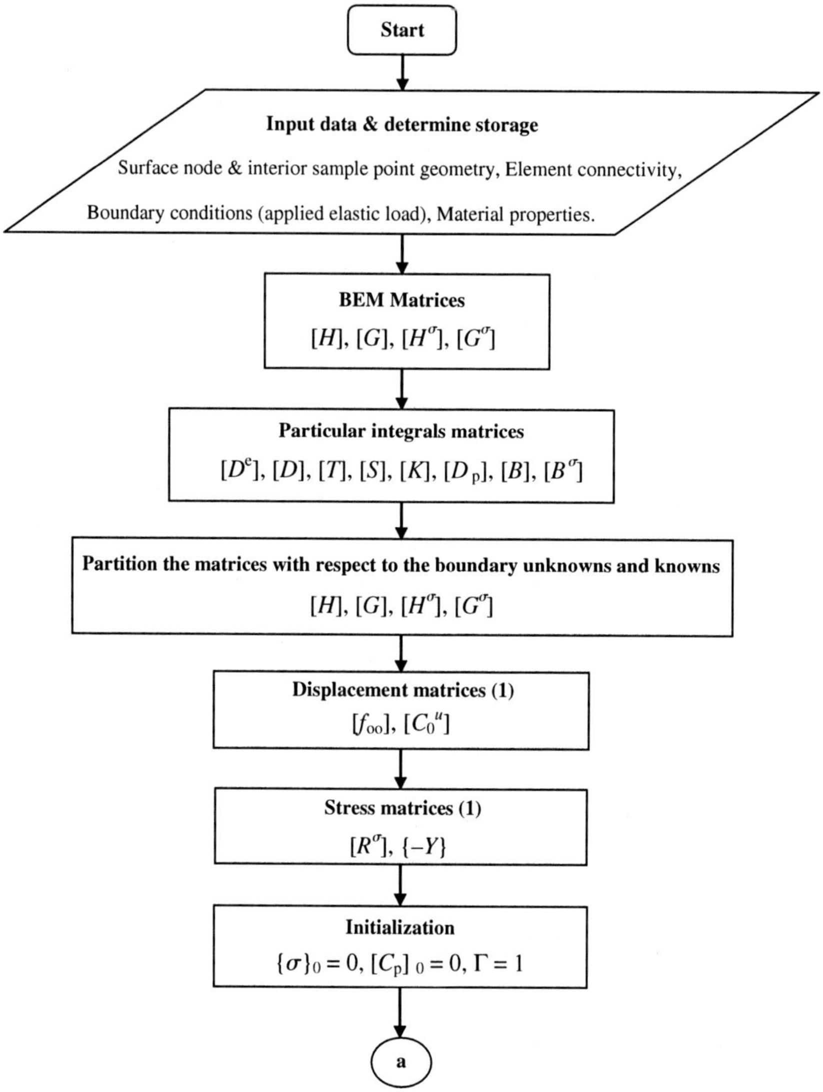

Figure D.1: Flowchart for elastoplastic analysis using BE-MP

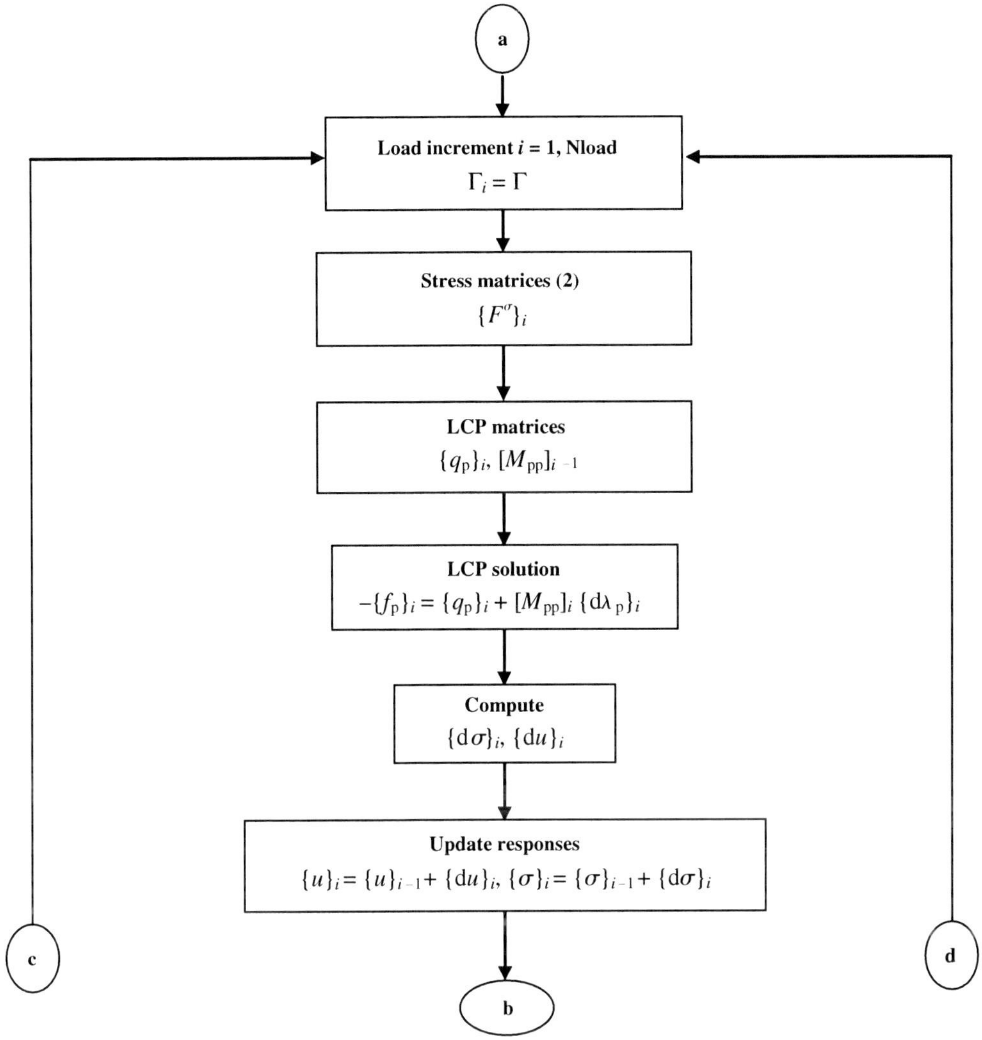

Figure D.1: *Continued.*

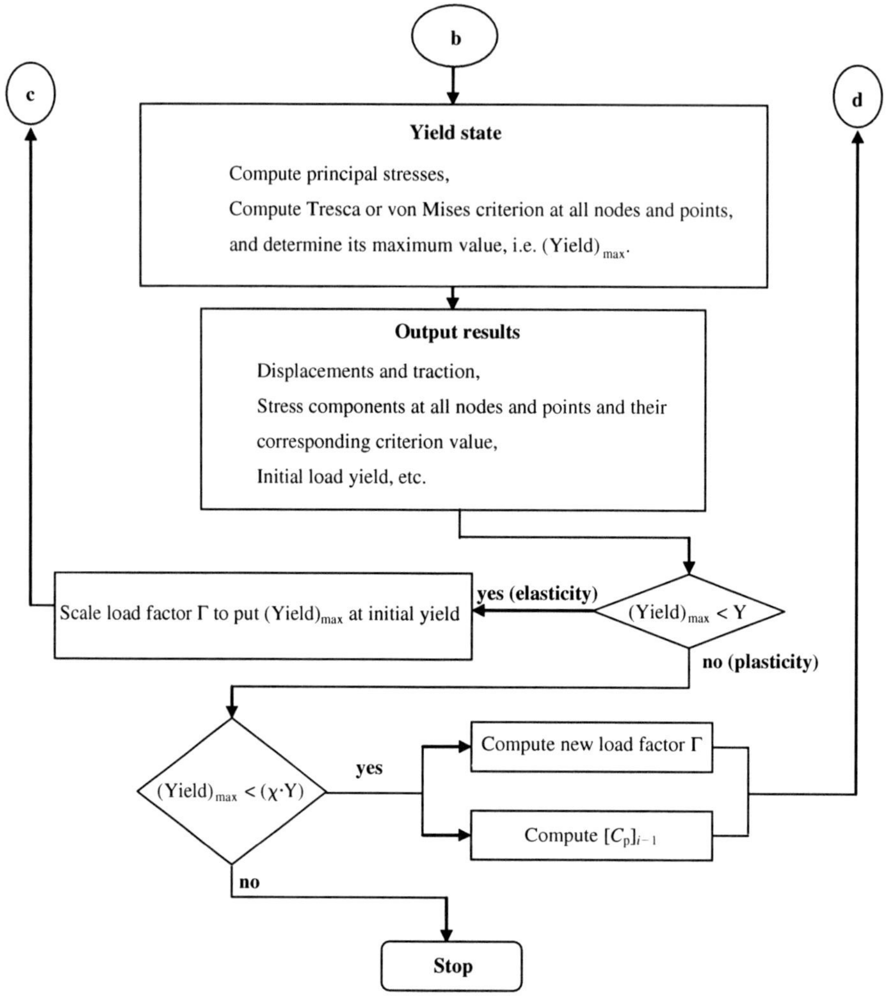

Figure D.1: *Continued.*

Bibliography

Abdul-Mihsein, M.J., Becker, A.A. & Parker, A.P., A boundary integral equation method for axisymmetric elastic contact problems. *Comput. Struct.*, **23(6)**, pp. 787–793, 1986.

Alart, P. & Curnier, A., A mixed formulation for frictional contact problems prone to Newton like solution methods. *Comput. Meth. Appl. Mech. Engrg.*, **92**, pp. 353–375, 1991.

Alart, P., A simple contact algorithm applied to large sliding and anisotropic friction. *Proceedings of the Contact Mechanics International Symposium*, ed. A. Curnier, PPUR, pp. 321–336, 1992.

Alcantud, M., Doblaré, M., Gracia, L. & Dominguez, J., Analysis of the contact problem between elastoplastic 2-D bodies by means of the BEM. *Boundary Elements in Mechanical and Electrical Engineering*, eds. C.A. Brebbia & A. Chaudouet-Miranda, Computational Mechanics Publications: Southampton, pp. 15–30, 1990.

Aleynikov, S.M., *Boundary Element Method in Contact Problems for Elastic Spatial & Nonhomogeneous Bases* (3d-Problems of Numerical Geotechnique), Publishing House of Civil Engineering Universities Association, Moscow, 2000.

Alexandrov, V.M. & Pozharskii, D.A., *Three-Dimensional Contact Problems*, Kluwer Academic Publisher: Dordrecht, 2002.

Aliabadi, M.H. & Brebbia, C.A., (eds). Computational methods in contact mechanics I. *Proceeding of the First International Conference*. Southampton, UK, WIT Press: UK, 1993.

Aliabadi, M.H. & Alessandri, C., (eds). Computational methods in contact mechanics II. *Proceeding of the Second International Conference*. Ferrara, Italy, WIT Press: UK, 1995.

Amontons, G., *Sur l'origine de la résistance dans les machines*, Mémoires de l'Académie Royale, pp. 206–222, 1699.

Andersson, T., Fredriksson, B. & Persson, B.G.A., The boundary element method applied to two-dimensional contact problems. *New Developments in Boundary Element Methods*, ed. C.A. Brebbia, pp. 247–263, 1980.

Andersson, T., The boundary element method applied to two-dimensional contact problems with friction. *Boundary Element Methods, Proceedings of the Third International Seminar*, ed. C.A. Brebbia, Springer: Berlin, pp. 239–258, 1981.

Andersson, T., The second generation boundary element contact problem. *Boundary Element Methods in Engineering, Proceedings of the Fourth International Seminar*, ed. C.A. Brebbia, Springer: Berlin, pp. 409–427, 1982.

Antes, H. & Panagiotopoulos, P.D., *The Boundary Integral Approach to Static & Dynamic Contact Problems: Equality and Inequality Methods*, Birkhauser: Basel, 1992.

Bathe, K.J. & Chaudhary, A.A., A solution method for planner and axisymmetric contact problems. *Int. J. Numer. Meth. Engrg.*, **21**, pp. 65–88, 1985.

Batoz, J.L. & Dhatt, G., *Modélisation des structures par éléments finis*, Malonie S.A. et Les Presses de l'Université Laval, Paris et Québec, 1990.

Bazaraa, M., Sherali, H.D. & Shetty, C.M., *Nonlinear Programming, Theory and Application*, Second Edition, John Wiley and Sons: New York, 1999.

Becker, A.A., *The Boundary Element Method in Engineering: A Complete Course*, McGraw-Hill Book Company: UK, 1992.

Beeching, R. & Nicholls, W., A theoretical discussion of pitting failures in gears. *Proc. Inst. Mech. Eng.*, **158**, pp. 317–326, 1948.

Belytschko, T., Liu, T.W. & Moran, B., *Nonlinear Finite Elements for Continua and Structures*, John Wiley & Sons: New York, 2000.

Bjorklund, S. & Andersson, S., A numerical method for real elastic contacts subjected to normal and tangential loading. *Wear*, **179**, pp. 117–122, 1994.

Bjorkman, G., Klarbring, A., Sjodin, B., Larsson, T. & Ronnqvist, M., Sequential quadratic programming for non-linear elastic contact problems. *Int. J. Numer. Meth. Engrg.*, **38**, pp. 137–165, 1995.

Blazques, A., Paris, F., Canas, J. & Garrido, J.A., An algorithm for frictionless contact problems with non-conforming discretizations using BEM. *Boundary Elements XIV*, Vol. 2: *Stress Analysis and Computational Aspects*, eds. C.A. Brebbia, J. Dominguez & F. Paris, Computational Mechanics Publications: Southampton, pp. 409–420, 1992.

Brebbia, C.A., *The Boundary Element Method for Engineers*, Pentech Press: London, 1978.

Brebbia, C.A. & Dominguez, J., *Boundary Elements: An Introductory Course*, McGraw-Hill Book Company: Toronto, 1989.

Brebbia, C.A., (ed). Computational methods in contact mechanics VI. *Proceeding of the Sixth International Conference*, Crete, Greece, WIT Press: UK, 2003.

Carter, R.W., On the action of locomotive driving wheel. *Proc. Roy. Soc. London*, **A 112**, pp. 151–157, 1926.

Cattaneo, C., Sul contatto di due corpi elastici: distribuzione locale degli sforzi. *Accademia dei Lincei, Rendiconti*, Series 6, fol. **27**, Part I: 342–348, Part II: 434–436, Part III: 474–478, 1938.

Chan, S.K. & Tuba, I.S., A finite element method for contact problems of solid bodies, Part I. Theory & Validation. *Int. J. Mech. Sci.*, **13**, pp. 615–625, 1971.

Chaudhary, A.B. & Bathe, K.J., A solution method for static and dynamic analysis of three-dimensional contact problems with friction. *Comput. Struct.*, **24(6)**, pp. 855–873, 1986.

Chen, W.F. & Saleeb, A.F., *Constitutive Equations for Engineering Materials*, Vol. I, II, Elsevier: Amsterdam, 1994.

Chen, W.H. & Yen, J.T., Finite element analysis of finite deformation contact problems with friction. *Comput. Struct.*, **29(3)**, pp. 423–436, 1988.

Cheng, J.H. & Kikuchi, N., An analysis of metal forming progress using large deformation elastic-plastic formulations. *Comput. Meth. Appl. Mech. Engrg.*, **49**, pp. 71–108, 1985.

Ciavarella, M., The generalized cattaneo partial slip plane contact problem, Part I: Theory. *Int. J. Solids Struct.*, **35(18)**, pp. 2349–2362, 1998a.

Ciavarella, M., The generalized cattaneo partial slip plane contact problem, Part I: Examples. *Int. J. Solids Struct.*, **35(18)**, pp. 2363–2378, 1998b.

Ciavarella, M., Tangential loading of general three-dimensional contacts. *J. Appl. Mech.*, **65**, pp. 998–1003, 1998c.

Cottle, R.W. & Dantzig, G.B., Complementary pivot theory of mathematical programming. *Linear Algebra Appl.*, **1**, pp. 103–125, 1968.

Cottle, R.W., Pang, J.S. & Stone, R.E., *The Linear Complementarity Problem*, Academic Press: Boston, 1992.

Coulomb, C.A., *Théorie des machines simple*, Mémoires de Mathématique et de Physique de l'Académie Royale, pp. 161–342, 1785.

Crisfield, M.A., *Non-linear Finite Element Analysis of Solids and Structures*, Vol. 2, John Wiley & Sons: New York, 1997.

Crisfield, M.A., *Non-linear Finite Element Analysis of Solids and Structures*, Vol. 1, Second Edition, John Wiley & Sons: New York, 2000.

Curnier, A., A Theory of friction. *Int. J. Solids Struct.*, **20**, pp. 637–647, 1984.

Curnier, A., Unilateral contact: mechanical modelling. *New Developments in Contact Problems*, eds. P. Wriggers & P. Panagiotopoulos, CISM, Courses and Lectures No. 384, Springer-Verlag: Berlin, pp. 1–54, 1999.

Dandekar, B.W. & Conant, R.J., Numerical analysis of elastic contact problems using the boundary integral equation method. Part 1: Theory. *Int. J. Numer. Meth. Engrg.*, **33**, pp. 1513–1522, 1992a.

Dandekar, B.W. & Conant, R.J., Numerical analysis of elastic contact problems using the boundary integral equation method. Part 2: Results. *Int. J. Numer. Meth. Engrg.*, **33**, pp. 1523–1535, 1992b.

Dantzig, G.B. & Cottle, R.W., Positive Semi-definite Programming, ORC 63–18 (RR), Operations Research Center, University of California, Berkeley, 1963.

de Pater, A.D., On the reciprocal pressure between two elastic bodies. The Mechanics of Contact between Deformable Bodies, *Proceedings of the Symposium of the International Union of Theoretical and Applied Mechanics*, eds. A.D. de Pater & J.J. Kalker, Delft University Press: Enchede, Netherlands, pp. 29–75, 1975.

de Pater, A.D. & Kalker, J.J., (eds). The mechanics of contact between deformable bodies. *Proceedings of the Symposium of the International Union of Theoretical and Applied Mechanics*, Delft University Press: Enchede, Netherlands, 1975.

Deresiewicz, H., Oblique contact of nonspherical elastic bodies. *J. Appl. Mech.*, pp. 623–624, 1957.

Deresiewicz, H., Mechanics of granular matter. *Adv. Appl. Mech.*, **5**, pp. 233–306, 1958.

Deresiewicz, H., A note on second-order hertz contact. *J. Appl. Mech.*, pp. 141–142, 1961.

Deresiewicz, H., Bodies in contact with applications to granular media, *in* Mindlin, R.D. and Applied Mechanics, ed. G. Herrmann, Pergamon Press Inc., pp. 105–147, 1974.

Dominguez, J. & Brebbia, C.A., (eds). Computational methods in contact mechanics V. *Proceeding of the Fifth International Conference*, Seville, Spain, WIT Press: UK, 2001.

Dundurs, J. & Stippes, M., Role of elastic constants in certain contact problems. *J. Appl. Mech.*, pp. 965–970, 1970.

Duvaut, G. & Lions, J.L., *Les inéquations en mécanique et en physique*, Dunod, Paris, 1972.

Endahl, N., On the finite element solution of the elastoplastic axisymmetric hertz contact problem. *Comput. Struct.*, **24(4)**, pp. 517–523, 1986.

Fabrikant, V.I., *Applications of Potential Theory in Mechanics*, Kluwer Academic Publishers: Netherlands, 1989.

Fabrikant, V.I., *Mixed Boundary Value Problems of Potential Theory and Their Applications in Engineering*, Kluwer Academic Publishers: Netherlands, 1991.

Faraji, A., Cardou, A. & Gakwaya, A., Effect of arbitrarily directed tangential force on elastic contacting bodies. *J. Engrg. Mech., ASCE*, **125(11)**, pp. 1324–1326, 1999.

Faraji, A., Cardou, A. & Gakwaya, A., A boundary element-mathematical programming method for solving elastoplastic contact problems. *Computational Methods in Contact Mechanics VI*, ed. C.A. Brebbia, WIT Press: Southampton, pp. 35–45, 2003.

Faraji, A., Gakwaya, A. & Cardou, A., A boundary element-mathematical programming method for solving elastic contact problems. *6eme Colloque National en Calcul des Structures*, eds. M. Potier-Ferry, M. Bonnet & A. Bignonnet, École Polytechnique (UMR 7649), France, pp. 329–336, 2003a.

Faraji, A., Gakwaya, A. & Cardou, A., A boundary element-mathematical programming method for solving elastoplastic problems. *Boundary Elements* XXV, eds. C.A. Brebbia, D. Poljak & V. Roje, WIT Press: Southampton, pp. 93–102, 2003b.

Fichera, G., Sul problema elastostatico di Signorini con ambigue condizioni al contorno, *Atti Acc. Nazionale Lincei*, S. VIII, p. 34, 1963.

Fichera, G., Problemi elastostatici con vincoli uni-laterali: il problema di Signorini con ambigue condizioni al contorno, *Mem, Accad. Naz. Lincei*, S. VIII, Vol. VII, Sez. I, 5, pp. 91–140, 1964.

Finlayson, B.A., *The Method of Weighted Residuals and Variational Principles*, Academic Press, 1972.

Fischer-Cripps, A.C., *Introduction to Contact Mechanics*, Springer: Berlin, 2000.

Fischer-Cripps, A.C., *Nanoindentation*, Springer-Verlag: New York, 2002.

Foces, A., Garrido, J.A. & Paris, F., Three-dimensional contact using BEM. *Computational Methods in Contact Mechanics*, eds. M.H. Aliabadi & C.A. Brebbia, Computational Mechanics Publications: Southampton, pp. 191–232, 1993.

Follansbee, P.S. & Sinclair, G.B., Quasi-static normal indentation of an elasto-plastic half-space by a rigid sphere I: analysis. *Int. J. Solids Struct.*, **20(1)**, pp. 81–91, 1984.

Francavilla, A. & Zienkiewicz, C.O., A note on numerical computation of elastic contact problems. *Int. J. Numer. Meth. in Engrg.*, **9**, pp. 913–924, 1975.

Fredricksson, B., Finite element solution of surface nonlinearities in structural mechanics with special emphasis to contact and fracture mechanics problems. *Comput. Struct.*, **6**, pp. 281–290, 1976.

Gakwaya, A., *Développments et applications de la méthode des élément finis de frontière à la mécanique des solides*, Thèse de doctorat, Faculté des Sciences et de Génie, Département de genie mécanique, Université Laval, 1983.

Gakwaya, A., Lambert, D. & Cardou, A., A boundary element and mathematical programming approach for frictional contact problems. *Comput. Struct.*, **42(3)**, 341–353, 1992.

Galin, L.A., *Contact Problems in the Theory of Elasticity*, Trans. from Russian by Moss, H., ed., I.N. Sneddon, Dept. of Mathematics, North Carolina State College, 1953–1961.

Galin, L.A., *Contact Problems of the Theory of Elasticity and Viscoelasticity*, Izdatelstvo Nauka, Moscow (in Russian), 1980.

Gaul, L. & Brebbia, C.A., (eds). Computational Methods in Contact Mechanics IV. *Proceeding of the Fourth International Conference*, Stuttgart, Germany, WIT Press: UK, 1999.

Gladwell, G.M.L., *Contact Problems in the Classical Theory of Elasticity*, Sijthoff and Noordhoff, 1980.

Goodman, L.E., Contact stress analysis of normally loaded rough spheres. *J. Appl. Mech.*, pp. 515–522, 1962.

Goodman, L.E., Development of the classical theory of contact stress. *Contact Mechanics and Wear of Rail/Wheel Systems*, eds. J. Kalousek, R.V. Dukkipati & G.M.L. Gladwell, University of Waterloo Press, pp. 1–10, 1982.

Goryacheva, I.G., *Contact Mechanics in Tribology*, Kluwer Academic Publishers: Netherlands, 1998.

Guiggiani, M. & Gigante, A., A general algorithm for multidimensional cauchy principal value integrals in the boundary element method. *J. Appl. Mech.*, **57**, pp. 906–915, 1990.

Guiggiani, M., Krishnasamy, G., Rudolphi, T.J. & Rizzo, F.J., A general algorithm for the numerical solution of hypersingular boundary integral equations. *J. Appl. Mech.*, **59**, pp. 604–614, 1992.

Gun, H., An efficient implementation of the boundary integral equation method in elastoplastic stress analysis of frictional contact problems. *Boundary Elements XXI*, eds. C.A. Brebbia & H. Power, WIT Press: Southampton, pp. 85–97, 1999.

Hardy, C., Baronet, C.N. & Tordion, G.V., The elasto-plastic indentation of a half-space by a rigid sphere. *Int. J. Numer. Meth. Engrg.*, **3**, pp. 451–462, 1971.

Harris, T.A., *Rolling Bearing Analysis*, John Wiley & Sons, 1966.

Hartmann, F., Computing the C-matrix in non-smooth boundary points. *New Developments in Boundary Element Methods: Proceedings of the Second International Seminar*, ed. C.A. Brebbia, London, pp. 367–379, 1980.

Heege, A. & Alart, P., A frictional contact element for strongly curved contact problems. *Int. J. Numer. Meth. Engrg.*, **39**, pp. 165–184, 1996.

Heems, M., Lagarde, F., Courtel, R. & Sorin, P., Sur les contraintes au contact de deux cylindres à axes parallèles chargés normalement et tangentiellement. *Comptes-rendus de l'Acad. des Sci.*, pp. 3293–3296, 1963.

Henry, Jr., D.P. & Banerjee, P.K., A new BEM formulation for two- and three-dimensional elastoplasticity using particular integrals. *Int. J. Numer. Meth. Engrg.*, **26**, pp. 2079–2096, 1988.

Hertz, H., *Miscellaneous Papers on the Contact of Elastic Solids*, Trans. by Jones, D.E., Macmillan and Co. Ltd.: London, 1896a.

Hertz, H., *Miscellaneous Papers on the Contact of Elastic Solids and on Hardness*, Trans. by Jones, D.E., Macmillan and Co. Ltd.: London, 1986b.

Hibbitt, Karlsson and Sorensen, Inc., *ABAQUS/Standard, User's Manual (Version 6.2)*, Vols I & II, 2001.

Hill, R., *The Mathematical Theory of Plasticity*, Oxford, 1983.

Hills, D.A., Nowell, D. & Sackfield, A., *Mechanics of Elastic Contacts*, Butterworth-Heinemann Ltd, 1993.

Hills, D.A. & Nowell, D., *Mechanics of Fretting Fatigue*, Kluwer Academic Publishers, 1994.

Huesmann, A. & Kuhn, G., Automatic load incrementation technique for plane elastoplastic frictional contact problems using boundary element method. *Comput. Struct.*, **56(5)**, pp. 733–744, 1995.

Jacobson, B. & Kalker, J.J., *Rolling Contact Phenomena*, Springer, 2000.

Jäger, J., Contact with friction of elastically similar bodies, in *AMD* Vol. 205, *Impact, Waves, and Fracture*, ASME, eds. R.C. Batra, A.K. Mal & G.P. Macsithigh, pp. 129–152, 1995.

Jäger, J., A new principle in contact mechanics. *J. Tribology*, **120**, pp. 677–684, 1998.

Jäger, J., Tangential loading of general three-dimensional contacts. *J. Appl. Mech.*, **66**, pp. 1048–1050, 1999.

Jäger, J., *New Solutions in Contact Mechanics*, WIT Press: UK, 2004.

Johnson, K.L., Surface interaction between elastically loaded bodies under tangential forces. *Proc. Roy. Soc. London*, **A 230**, pp. 531–549, 1955.

Johnson, K.L., Inelastic contact: plastic flow and shakedown. *Contact Mechanics and Wear of Rail/Wheel Systems*, eds. J. Kalousek, R.V. Dukkipati & G.M.L. Gladwell, University of Waterloo Press, pp. 79–101, 1982a.

Johnson, K.L., One hundred years of hertz contact. *Proc. Inst. Mech. Engrs.*, **196**, pp. 363–378, 1982b.

Johnson, K.L., *Contact Mechanics*, Cambridge University Press, 1985.

Joo, J.W. & Kwak, B.M., Analysis and applications of elasto-plastic contact problems considering large deformation. *Comput. Struct.*, **24(6)**, pp. 953–961, 1986.

Kachanov, L.M., *Fundamentals of the Theory of Plasticity*, North-Holland Publishing Company: Amsterdam, 1971.

Kalker, J.J., A survey of the mechanics of contact between solid bodies. *ZAMM*, **57**, pp. 3–17, 1977.

Kalker, J.J., *Three-Dimensional Elastic Bodies In Rolling Contact*, Kluwer Academic Publishers, 1990.

Kane, J.H., *Boundary Element Analysis in Engineering Continuum Mechanics*, Prentice Hall: New Jersey, 1994.

Karami, G., *A Boundary Element Method for Two-Dimensional Contact Problems*, Springer-Verlag: Berlin, 1989.

Karami, G., Boundary element analysis of elasto-plastic contact problems. *Comput. Struct.*, **41(5)**, pp. 927–935, 1991.

Keer, L.M., Ahmadi, N. & Mura, T., Tangential loading of elastic bodies in contact. *Comput. Struct.*, **19(1–2)**, pp. 93–101, 1984.

Keer, L.M. & Goodman, L.E., Tangential loading of two bodies in contact. *J. Appl. Mech.*, pp. 513–514, 1976.

Khludnev, A.M. & Sokolowski, J., *Modelling and Control in Solid Mechanics*, Birkhauser Verlag: Berlin, 1997.

Kikuchi, N., A smoothing technique for reduced integration penalty methods in contact problems. *Int. J. Numer. Meth. Engrg.*, **18**, pp. 343–350, 1982.

Kikuchi, N. & Oden, J.T., Contact problems in elastostatics. *Finite Element: Special Problems in Solid Mechanics*, eds. J.T. Oden & G.F. Carey, Vol. 5, Chap. 4, 1984.

Kikuchi, N. & Oden, J.T., *Contact Problems in Elasticity: A Study of Variational Inequalities and Finite Element Methods*, SIAM, Philadelphia, 1988.

Kinderlehrer, D. & Stampacchia, G., *An Introduction to Variational Inequalities and their Applications*, Academic Press: New York, 1980.

Klarbring, A., A mathematical programming approach to three-dimensional contact problems with friction. *Comput. Meth. Appl. Mech. Engrg.*, **58**, pp. 175–200, 1986a.

Klarbring, A., General contact boundary conditions and the analysis of frictional systems. *Int. J. Solids Struct.*, **22(12)**, pp. 1377–1398, 1986b.

Klarbring, A., On discrete and discretized non-linear elastic structures in unilateral contact (stability, uniqueness and variational principles). *Int. J. Solids Struct.*, **24(5)**, pp. 459–479, 1988.

Klarbring, A., Derivation and analysis of rate boundary-value problems of frictional contact. *Eur. J. Mech. A/Solids*, **9(1)**, pp. 53–85, 1990.

Klarbring, A., Mathematical programming and augmented lagrangian methods for frictional contact problems. *Proceedings of the Contact Mechanics International Symposium*, ed. A. Curnier, PPUR, pp. 409–422, 1992.

Klarbring, A., Mathematical programming in contact problems. *Computational Methods in Contact Mechanics*, eds. M.H. Aliabadi & C.A. Brebbia, Computational Mechanics Publications: Southampton, pp. 233–263, 1993.

Klarbring, A., Contact, friction, discrete mechanical structures and mathematical programming. *New Development in Contact Problems*, eds. P. Wriggers & P. Panagiotopoulos, CISM, Courses and Lectures No. 384, Springer-Verlag, pp. 55–100, 1999.

Klarbring, A. & Bjorkman, G., A mathematical programming approach to contact problems with friction and varying contact surfaces. *Comput. Struct.*, **30(5)**, pp. 1185–1198, 1988.

Klarbring, A. & Bjorkman, G., Solution of large displacement contact problems with friction using Newton's method for generalized equations. *Int. J. Numer. Meth. Engrg.*, **34**, pp. 249–269, 1992.

Kong, X.A., Gakwaya, A., Cardou, A. & Cloutier, L., A numerical solution of general frictional contact problems by the direct boundary element and mathematical programming approach. *Comput. Struct.*, **45(1)**, pp. 95–112, 1992.

Kong, X.A., Gakwaya, A., Cardou, A. & Cloutier, L., An elasto-plastic contact BEM by particular integrals and mathematical programming. *Eur. J. Mech. A/Solids*, **12(4)**, pp. 537–566, 1993.

Kragelsky, I.V., Dobychin, M.N. & Kombalov,V.S., *Friction and Wear Calculation Methods*, Pergamon Press: Oxford, 1982.

Kuich, G., Applications of the boundary element method to contact problems, in *BETECH*, ed. C.A. Brebbia, Computational Mechanics Publications: Southampton, pp. 499–519, 1986.

Kwak, B.M., A complementarity problem formulation of three-dimensional frictional contact problems. *J. Appl. Mech.*, **58**, pp. 134–140, 1991.

Kwak, B.M. & Lee, S.S., A complementarity problem formulation for two-dimensional frictional contact problems. *Comput. Struct.*, **28(4)**, pp. 469–480, 1988.

Laursen, T.A., *Computational Contact and Impact Mechanics*, Springer, 2002.

Lee, B.C. & Kwak, B.M., A computational method for elastic-plastic contact problems. *Comput. Struct.*, **18(5)**, pp. 757–765, 1984.

Lee, K., An efficient solution method for frictional contact problems. *Comput. Struct.*, **32(1)**, pp. 1–11, 1989.

Lemaitre, J. & Chaboche, J.L., *Mécanique des matériaux solides,* Bordas, Paris, 1988.

Lemke, C.E. & Howson, J.T., Equilibrium points of bimatrix games. *SIAM J. Appl. Math.*, **12(20)**, pp. 413–423, 1964.

Lemke, C.E., On complementary pivot theory. *Mathematics of the Decision Sciences, Part I* (Am. Math. Soc., Providence, RI), eds. G.B. Dantzig & A.F. Veinott, pp. 95–114, 1968.

Li, M., Sha, D. & Tamma, K.K., Linear complementary formulation involving frictional contact for elasto-Plastic deformable bodies. *J. Appl. Mech.*, **64**, pp. 80–89, 1997.

Ling, F.F., *Surface Mechanics*, John Wiley & Sons, 1973.

Ling, F.F., Lai, W.M. & Lucca, D.A., *Fundamentals of Surface Mechanics with Applications*, Second Edition, Springer, 2002.

Love, A.E.H., *A Treatise on The Mathematical Theory of Elasticity*, Dover Publications: New York, 1944.

Lubkin, J.L., The torsion of elastic spheres in contact. *J. Appl. Mech.*, pp. 183–187, 1951.

Ludema, K.C., *Friction, Wear, Lubrication: A Textbook in Tribology*, CRC Press: Florida, 1996.

Luré, A.I., *Three-Dimensional Problems of the Theory of Elasticity*, Interscience: New York, 1964.

Madhusudana, C.V., *Thermal Contact Conductance*, Springer: Berlin, 1996.

Maier, G. & Munro, J., Mathematical programming applications to engineering plastic analysis. *Appl. Mech. Rev.*, **35(12)**, pp. 1631–1643, 1982.

Maier, G. & Novati, G., Elastic-plastic boundary element analysis as a linear complementarity problem. *Appl. Math. Model.*, **7**, pp. 74–82, 1983.

Maier, G. & Polizzotto, C., A Galerkin approach to boundary element elastoplasticity analysis. *Comput. Meth. Appl. Mech. Engrg.*, **60**, pp. 175–194, 1987.

Man, K.W., *Contact Mechanics Using Boundary Elements*, Computational Mechanics Publications: Southapmton, 1994.

Man, K.W., Aliabadi, M.H. & Rooke, D.P., BEM frictional contact analysis: load incremental technique. *Comput. Struct.*, **47(6)**, pp. 893–905, 1991.

Martin, J.B., *Plasticity: Fundamentals and General Results*, The MIT Press: Cambridge, 1975.
Martin, D., Aliabadi, M.H. & Leitao, V.M.A., Application of BEM to elastoplastic contact problems. *Proceedings of the 16th International Conference On Boundary Element Methods*, Southampton, 1994.
Maugis, D., *Contact, Adhesion and Rupture of Elastic Solids*, Springer, 2000.
M'Ewen, E., Stresses in elastic cylinders in contact along generatrix (including the effect of tangential friction). *Phil. Mag.*, **40**, pp. 455–459, 1949.
Mindlin, R.D., Compliance of elastic bodies in contact. *J. Appl. Mech.*, pp. 259–268, 1949.
Mindlin, R.D., Mason, E.P., Osmer, T.F. & Deresiewicz, H., Effects of an oscillating tangential force on the contact surfaces of elastic spheres. *Proceedings of the First National Congress of Applied Mechanics*, pp. 203–208, 1951.
Mohammadi, S., *Discontinuum Mechanics Using Finite and Discrete Elements*, WIT Press: UK, 2003.
Mroz, Z. & Stupkiewicz, S., Constitutive modeling of slip and wear in elastic, frictional contact. *Proceedings of the Contact Mechanics International Symposium*, ed. A. Curnier, PPUR, pp. 133–156, 1992.
Murty, K.G., *Linear Complementarity, Linear and Nonlinear Programming*, Heldermann Verlag: Berlin, 1988.
Muskhelishvili, N.I., *Some Basic Problems of the Mathematical Theory of Elasticity*, Trans. by Radok, J.R.M., P. Noordhoff Ltd, 1963.
Nadai, A., *Theory of Flow and Fracture of Solids*, Vol. II, McGraw-Hill: Toronto, 1963.
Nardini, D. & Brebbia, C.A., A new approach to free vibration analysis using boundary elements. *Boundary Elements*, Vol. IV, ed. C.A. Brebbia, Springer-Verlag, pp. 312–348, 1982.
Nayak, G.C. & Zienkiewicz, O.C., Convenient form of stress invariants for plasticity. *J. Struct. ASCE*, **98(ST 4)**, pp. 949–953, 1972.
Oden, J.T. & Carey, G.F., *Finite Elements: Special Problems in Solid Mechanics*, Vol. V, Prentice-Hall: Englewood Cliffs, NJ, 1984.
Oden, J.T. & Kikuchi, N., Finite element methods for constrained problems in elasticity. *Int. J. Numer. Meth. Engrg.*, **18**, pp. 701–725, 1982.
Oden, J.T. & Martins, J.A.C., Models and computational methods for dynamic friction phenomena. *Comput. Meth. Appl. Mech. Engrg.*, **52**, pp. 527–634, 1985.
Oden, J.T. & Pires, E.B., Numerical analysis of certain contact problems in elasticity with non-classical friction laws. *Comput. Struct.*, **16(1–4)**, pp. 481–485, 1983.
Okamoto, N. & Nakazawa, M., Finite element incremental contact analysis with various frictional conditions. *Int. J. Numer. Meth. Engrg.*, **14**, pp. 337–357, 1979.
Olukoko, O.A., Becker, A.A. & Fenner, R.T., A new boundary element approach for contact problems with friction. *Int. J. Numer. Meth. Engrg.*, **36**, pp. 2625–2642, 1993.
Owen, D.R.J. & Hinton, E., *Finite Elements in Plasticity*, Pineridge Press Ltd.: Swansea, UK, 1980.
Panagiotopoulos, P.D., *Inequality Problems in Mechanics and Applications*, Birkhauser, 1985.
Paris, F. & Garrido, J.A., On the use of discontinuous elements in two-dimensional contact problems. *Boundary Elements VII, Proceedings of the Seventh International Conference*, eds. C.A. Brebbia & G. Maier, Computational Mechanics Publications: Southampton 13–17, pp. 33–39, 1985.
Paris, F., Blazquez, A. & Canas, J., Contact problems with nonconforming discretizations using boundary element method. *Comput. Struct.*, **57(5)**, pp. 829–839, 1995.
Paul, B. & Hashemi, J., Contact pressures on closely conforming elastic bodies. *Solid Contact and Lubrication*, eds. H.S. Cheng & L.M. Keer, **39**, pp. 67–78, 1980.

Pfeiffer, F. & Glocker, C., *Multibody Dynamics with Unilateral Contacts*, John Wiley & Sons: New York, 1996.

Polizzotto, C. & Zito, M., BIEM-based variational principles for elastoplasticity with unilateral contact boundary conditions. *Engrg. Anal. Boundary Element*, **21**, pp. 329–338, 1998.

Poritsky, H., Stress and deflections of cylindrical bodies in contact with application to contact of gears and of locomotive wheels. *J. Appl. Mech.*, pp. 191–201, 1950.

Prasad, S.N. & Dasgupta, S., Effect of sliding friction on contact stresses in a rectangle compressed by rigid planes. *J. Appl. Mech., ASME*, **42**, pp. 656–662, 1975.

Raoof, M. & Hobbs, R.E., Tangential compliance of rough elastic bodies in contact. *J. Tribology*, **111**, pp. 726–729, 1989.

Raous, M., Chabrand, P. & Lebon, F., (eds). Numerical methods in mechanics of contact involving friction. *Journal de mécanique théorique et appliquée*, Special Issue, **7**, (Suppl. No. 1), 1988.

Raous, M., Quasistatic signorini problem with coulomb friction and coupling to adhesion. *New Development in Contact Problems*, eds. P. Wriggers & P. Panagiotopoulos, CISM, Courses and Lectures No. 384, Springer-Verlag, pp. 101–178, 1999.

Reddy, J.N. & Rasmussen, M.L., *Advanced Engineering Analysis*, John Wiley & Sons: Toronto, 1982.

Ravindran, A., A computer routine for quadratic and linear programming problems [H], *Commun. ACM*, **15(9)**, pp. 818–820, 1972.

Rekach, V.G., *Manual of the Theory of Elasticity*, Trans. from Russian by M. Konyaeva, Mir Publishers: Moscow, 1979.

Rizzo, F.J. & Shippy, D.J., A formulation and solution procedure for the general non-homogeneous elastic inclusion problem. *Int. J. Solids Struct.*, **4**, pp. 1161–1179, 1968.

Sachdeva, T.D. & Ramakrishnan, C.V., A finite element solution for the two-dimensional elastic contact problems with friction. *Int. J. Numer. Meth. Engrg.*, **17**, pp. 1257–1271, 1981.

Samartin, A. & Aliabadi, M.H., (eds). Computational methods in contact mechanics III. *Proceeding of the Third International Conference*, Madrid, Spain, WIT Press: UK, 1997.

Saran, M.J. & Taylor, R.L., An application of augmented lagrangian method for finite element simulation of large deformation frictional contact problems, CED Vol. 5/AMD Vol. 156, *Numerical Methods for Simulation of Industrial Metal Forming Processes*, ASME, pp. 19–26, 1992.

Seireg, A.A., *Friction and Lubrication in Mechanical Design*, Marcel Dekker: New York, 1998.

Sextro, W., *Dynamical Contact Problems with Friction Models, Methods, Experiments and Applications,* Springer-Verlag: New York, 2002.

Sha, D., Sun, H., Zhang, Z. & Yin, F., A variational inequality principle in solid mechanics and application in physically non-linear problems. *Commun. Appl. Numer. Meth.*, **6**, pp. 35–45, 1990.

Sha, D., Li, M. & Tamma, K.K., An effective linear complementary model involving general contact with coulomb friction, CED Vol. 5/AMD Vol. 156, *Numerical Methods for Simulation of Industrial Metal Forming Processes*, ASME, pp. 27–33, 1992.

Signorini, A., Sopra akune questioni di elastostatica, Atti della Societa Italiana per il Progresso delle Scienze, 1933.

Signorini, A., Questioni di elasticità nonlinearizzata e semi linearizzata. *Rend. Mat.*, **18**, pp. 95–139, 1959.

Simunovic, S. & Saigal, S., Frictionless contact with BEM using quadratic programming. *J. Appl. Mech.*, **118(9)**, pp. 1876–2540, 1992.

Simunovic, S. & Saigal, S., Quadratic programming contact formulation for elastic bodies using boundary element method. *AIAA J.*, **33(2)**, pp. 325–331, 1995a.

Simunovic, S. & Saigal, S., A linear programming formulation for incremental contact analysis. *Int. J. Numer. Meth. Engrg.*, **38**, pp. 2703–2725, 1995b.

Sinclair, G.B., Follansbee, P.S. & Johnson, K.L., Quasi-static normal indentation of an elasto-plastic half-space by a rigid sphere II. Results, *Int. J. Solids Struct.*, **21(8)**, pp. 865–888, 1985.

Singer, I. & Pollock, H.M., *Fundamentals of Friction: Macroscopic and Microscopic Processes*, Kluwer Academic Publishers, 1992.

Smith, J.O., Liu, C.K. & Ill, U., Stresses due to tangential and normal loads on an elastic solid with application to some contact stress problems. *J. Appl. Mech.*, pp. 157–166, 1953.

Sofonea, M. & Han, W., Quasistatic contact problems in viscoelasticity and viscoplasticity. *Studies in Advanced Mathematics*, Americal Mathematical Society: International Press, 2002.

Sofonea, M. & Viano, J.M., (eds). Mathematical modelling and numerical analysis in solid mechanics. *Int. J. Appl. Math. Comput. Sci.*, Special Issue, **12(1)**, 2002.

Spence, D.A., An eigenvalue problem for elastic contact with finite friction. *Proc. Camb. Philos. Soc.*, **73**, pp. 249–268, 1973.

Spence, D.A., The hertz contact problem with finite friction. *J. Elasticity*, **5**, pp. 297–319, 1975.

Spence, D.A., Frictional contact with transverse shear. *Q. J. Mech. Appl. Math.*, **2**, pp. 233–253, 1986.

Stavroulakis, G.E., Panagiotopoulos, P.D. & Al-Fahed, A.M., On the rigid body displacements and rotations in unilateral contact problems and applications. *Comput. Struct.*, **40(3)**, pp. 599–614, 1991.

Stolarski, T.A. & Tobe, S., *Rolling Contacts*, ASME Publication, 2001.

Stroud, A.H. & Secrest, D., *Gaussian Quadrature Formulas*, Prentice-Hall: 1966.

Takahashi, S., *Elastic Contact Analysis by Boundary Elements*, Springer-Verlag: Berlin, 1991.

Takahashi, S. & Brebbia, C.A., Analysis of contact problems in elastic bodies using a B.E.M. flexibility approach. *Boundary Elements X*, Vol. 3: *Stress Analysis*, ed., C.A. Brebbia, Computational Mechanics Publications: Southampton, pp. 353–379, 1988.

Tamma, K.K., Li, M. & Sha, D., Conjugate gradient based projection: a new explicit computational methodology for frictional contact problems, *Commun. Numer. Meth. Engrg.*, **10**, pp. 633–648, 1994.

Thomas, H.R. & Hoersch, V.A., Stresses due to the pressure of one elastic solid upon another, engineering experiment station, University of Illinois, *Bulletin No. 212*, pp. 1–54, 1930.

Torstenfelt, B., Contact problems with friction in general purpose finite element computer programs. *Comput. Struct.*, **16(1–4)**, pp. 487–493, 1983.

Torstenfelt, B., An automatic incrementation technique for contact problems with friction. *Comput. Struct.*, **19(3)**, pp. 393–400, 1984.

Trinkle, J.C., Pang, J.S., Sundarsky, S. & Lo, G., On dynamic multi-rigid-body contact problems with coulomb friction. *ZAMM*, **77(4)**, pp. 267–279, 1997.

Tsai, P. & Chen, W.H., Finite element analysis of elastoplastic contact problems with friction. *AIAA J.*, **24(2)**, pp. 344–346, 1986.

Tsuta, T. & Yamaji, S., Boundary element analysis of contact thermo-elastoplastic problems with creep and the numerical technique. *Boundary Elements, Proceedings of the Fifth International Conference*, eds. C.A. Brebbia, T.F. Futagami & M. Tanaka, Hiroshima, pp. 567–576, 1983.

Valliappan, S., Lee, I.K. & Boonlualohr, P., Non-linear analysis of contact problems. *Numerical Methods in Coupled Systems*, Chap. 8, eds. R.W. Lewis, P. Bettess & E. Hinton, John Wiley & Sons, pp. 231–253, 1984.

Vermeulen, P.J. & Johnson, K.L., Contact of nonspherical elastic bodies transmitting tangential forces. *J. Appl. Mech.*, pp. 338–340, 1964.

Vu-Quoc, L. & Zhang, X., An elasto-plastic contact force-displacement model in the normal direction: displacement-driven version. *Proc. Roy. Soc. London, Ser. A*, **455** (1991), pp. 4013–4044, 1999.

Wriggers, P., *Computational Contact Mechanics*, John Wiley & Sons, 2002.

Xu, B.Q. & Kong, X.A., A modified tree search algorithm for contact BEM-LCP approach, *Comput. Struct.*, **59(4)**, pp. 707–713, 1996.

Yamazaki, K. & Mori, M., Boundary element analysis of the elastic contact problem by a penalty function method. *Boundary Element Methods in Applied Mechanics*, eds. M. Tanaka & T.A. Crise, Pergamon Press, pp. 169–178, 1988.

Yamazaki, K., Penalty function method using BEM. *Computational Methods in Contact Mechanics*, eds. M.H. Aliabadi & C.A. Brebbia, Computational Mechanics Publications: Southampton, pp. 155–190, 1993.

Yamazaki, K., Sakamoto, J. & Takumi, S., Penalty method for three-dimensional elastic contact problems by boundary element method. *Comput. Struct.*, **52(5)**, pp. 895–903, 1994.

Yan, P., The effect of number of cycles on the critical transition boundary between fretting fatigue and fretting wear. *Wear*, **160**, pp. 279–289, 1993.

Zambelli, G. & Vincent, L., *Matériaux et contacts, une approche tribologique*, Presses polytechniques et universitaires romandes, Lausanne, 1998.

Zhong, W.X. & Sun, S.M., A finite element method for elasto-plastic structures and contact problems by parametric quadratic programming. *Int. J. Numer. Meth. Engrg.*, **26**, pp. 2723–2738, 1988.

Zhong, W.X. & Zhang, R.L., Parametric variational principles and their quadratic programming solutions in plasticity. *Comput. Struct.*, **30(4)**, pp. 887–896, 1988.

Zhong, W.X. & Wu, C.W., Elastic-plastic contacts using parametric quadratic programming. *Computational Methods in Contact Mechanics*, eds. M.H. Aliabadi & C.A. Brebbia, Computational Mechanics Publications: Southampton, pp. 305–352, 1993.

Zhong, Z.H., *Finite Element Procedures for Contact-Impact Problems*, Oxford University Press: Oxford, 1993.

Zhu, C. & Jin, Y., A finite element-mathematical programming method for elastoplastic problems based on the principle of virtual work. *Appl. Math. Mech. (English Version)*, **14(7)**, pp. 635–642, 1993.

Zhu, C., A finite element-mathematical programming method for elastoplastic contact problems with friction. *Finite Elements Anal. Design*, **20**, pp. 273–282, 1995a.

Zhu, C., A new boundary element-mathematical programming method for contact problems with friction. *Commun. Numer. Meth. Engrg.*, **11**, pp. 683–690, 1995b.